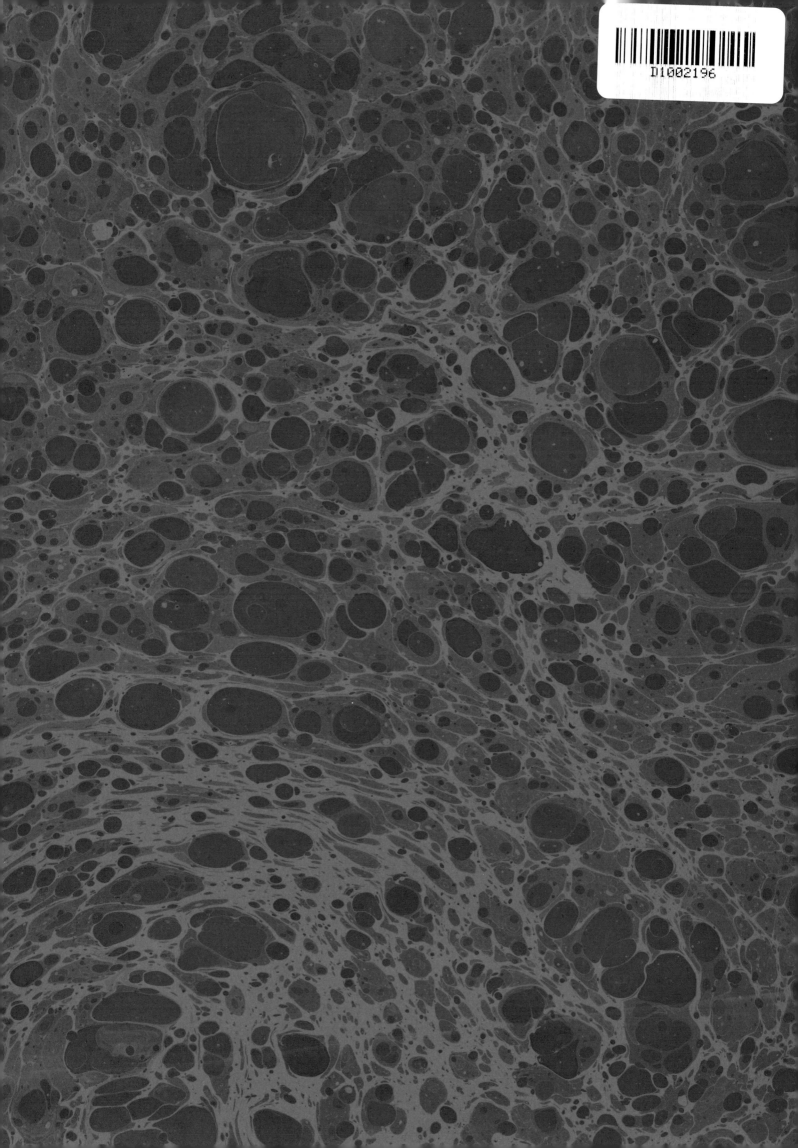

The Enchanted World

THE BOOK OF BEGINNINGS

The Enchanted World

THE BOOK OF BEGINNINGS

by the Editors of Time-Life Books

The Content

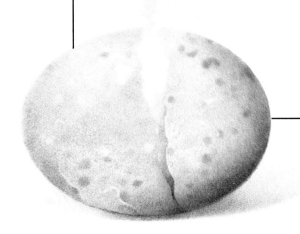

Chapter Three

The Dance of Life · 96

Time-Life Books · Alexandria, Virginia

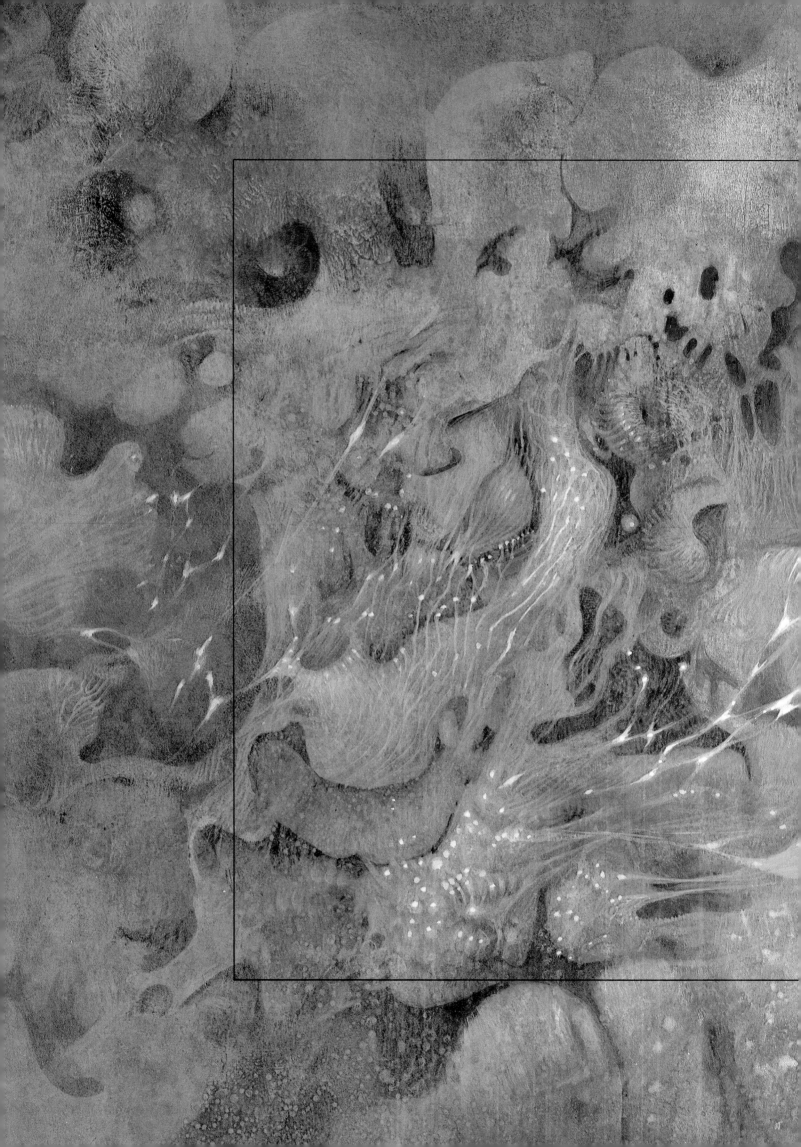

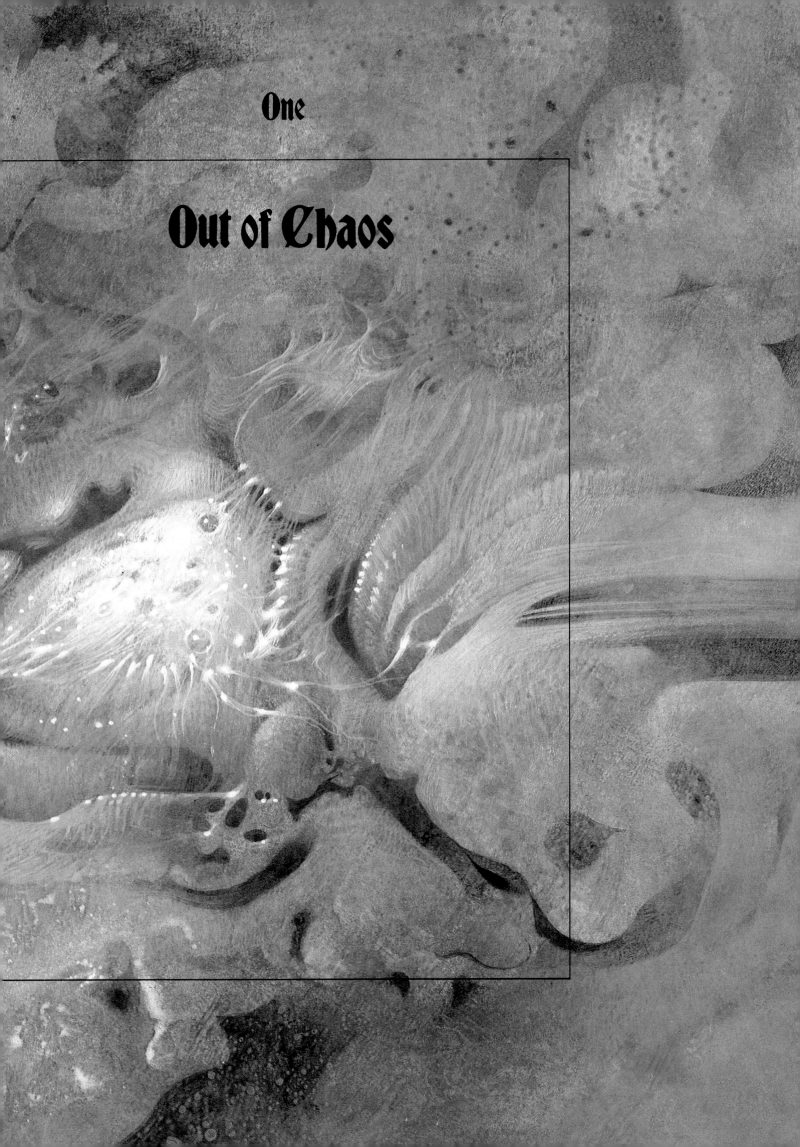

One

Out of Chaos

The Water-Mother

In the lands far to the north, in the lost ages when it seemed that the night lasted all winter long, the storytellers would gather around roaring fires to sing away the darkness. Old women, whirling spindles, told of life as it used to be in the days of their grandmothers' grandmothers. The healers and shamans, their fingers stained with the juices of potent herbs used in their mysterious work, recounted great deeds and sorrows from the morning of their race. As they passed round the beaker of mead, each bard would take it in turn to sing or speak an old story. Their memories were prodigious. The bards carried the entire history of their people, all their hard-won skills and hard-learned lessons, fastened up and secured against loss in a tightly woven net of rhythm and rhyme.

In the heavy stillness of the night, they shared their wisdom. And as the resin in the pine logs snapped and spat in the flames, or the great ice-sheet creaked and groaned on the black lake outside the door, someone would ask, inevitably, for the story behind all stories.

"Tell us again," they would whisper, "how everything began."

The question had many answers, none of them simple. Some came from distant places and strange peoples, carried in the bundled belongings of migrating tribes, the packs of peddlers or the baggage of marauding armies. Others were native to the place in which they were told, nearly as old as the land itself.

Perhaps, according to some stories, the world had simply grown from a celestial acorn into a vast, many-branched oak tree, whose sturdy limbs supported the earth and whose green foliage gave shelter to all living beings. Or it was said that the world was the work of a master craftsman, a mighty blacksmith as tall as the sky and as broad as the horizon, who stood sweating at an ancient forge heated by the primordial fire. Using his mysterious skills with hammer, tongs and bellows, he had slowly transformed molten chaos into cool creation.

Others told tales of a giant egg, of unexplained provenance, that hatched forth the earth and all the living things upon it. And those people who lived shivering in the high, cold places had their own vision of the beginning of the world, harking back to a time when everything was ice, and the world was created out of the body of a giant formed of frost.

Before the beginning of time, a goddess looked down from the galleries of her windswept palace, high in the Kingdom of the Air. She searched the clouds for a glimpse of the mysterious world below, not knowing that she herself was fated to play a part in its creation.

But those bards whose memories were longest spoke in awed voices of the Great Mother, giver of life. She was a shadowy figure, a queen of many names, remote as the stars but close as the generous earth underfoot. A loving source of life, she was also a dealer in death. Her terrible anger manifested itself in storms and plagues and cataclysm; she was capricious, all-powerful, quick to punish, magnanimous in her forgiveness. The earth itself may have been her body, the mountains her breasts, sending forth their nourishing milk as the streams that fed the sea. Spring and summer were her seasons of fecundity, the fall and winter her seasons of inviolable rest when she slept deeply, wrapped in unfathomable dreams, indifferent to the sufferings of her children.

Or perhaps she was always remote from them. Often she needed to be summoned and cajoled—by the hot, sweet smell of a blood sacrifice, or the worship of carefully aligned tall stones, faceless, wide-hipped statues and sacred fires.

Her visible manifestations were many. Sometimes she was a crone, wise and wrinkled; at other times she appeared as a snow-white sow or a dun-colored mare.

But one old Finnish chronicle remembered her as a young goddess who lived in a time before time. She had two names to mark the two phases of her shadowy existence—first as the Virgin of the Air, then later as the Water-Mother.

Her origins were mysterious. It was said that she was the daughter of the King of the Air. Nothing more was known of her father, or of his kingdom, beyond the fact that he had a palace. Floating in the empyrean, high above an endless expanse of water, it was hidden behind the curtain of the northern lights, walled by mist, roofed by rainbows. Its eight thousand chambers echoed with eight thousand kinds of emptiness, its glittering windows looked out on nothing and its corridors rolled on forever, meeting and branching, widening and narrowing, their far ends somewhere beyond the vanishing point.

This was the home of the Virgin of the Air. How she lived, what she fed on, what tongue she spoke in was a matter for conjecture. But it was known that she was as restless as the wind that scoured the shining halls of the palace. And one day she leaped—or perhaps fell, or might even have been pushed by some unseen, fateful force—through an enormous door or window that opened onto the void.

She plummeted through the bottomless caverns of cloud. If she screamed, it must have echoed through the universe, for she was a giantess of vast dimensions. But the wind currents caught her as if she were a weightless wisp, and carried her gently downward to the waiting ocean. As she fell slowly toward the whitecaps, the waves crested and reached up to catch her. For a while, the two elements tossed her back and forth between them. The sea lifted her up to receive the wind's caresses, then snatched her suddenly away.

But soon the game between the wind and the water grew deadly serious. The ocean surged up into angry billows; the wind whipped itself into a furious tempest and stirred the waves into a whirlpool.

The two opposing forces fought for her savagely, heedless of her frantic cries.

The maelstroms stilled, the roaring gusts subsided. There was nothing but silence. The giantess floated, no longer a virgin. Her body swelled with new life growing deep inside her, planted there by the violent, competitive courtship of storm and sea.

Now came a time of gestation. While she waited for this mysterious fruit to ripen she swam through the waters, and it was from this point in the tale that the bards began to call her the Water-Mother. Backward and forward she swam, finding nothing but ocean. Sometimes, to ease her burden, she turned over and floated on her back, gazing up toward her old home in the realm of the air. But she had fallen so far in her headlong dive that the palace was lost to view. If her father, the elusive King of the Air, knew of her predicament, he gave no sign.

As a goddess and an immortal the Water-Mother had no means or need to measure time. But it was recorded that she drifted there, with the waters supporting her massively swollen belly, for the equivalent of seven hundred years.

Year after year, century after century, she floated first east, then westward, then northward, then to the south. Sometimes she thrashed, restless and fretful. More often she drifted, deep in a trance, nourishing the child in her womb with her life's blood and her divine powers.

For all those years she waited, complete in herself, her energies and attention directed inward, contemplating the mysteries of quickening life. The moment came when she was no longer alone. Inexplicably, out of the nothingness or from some other, more populous sphere where deities danced and made magic, there came a winged messenger: a giant teal, boldly striped and speckled. The bird dipped low over the water, then rose and swooped again, searching urgently for a place to land.

The Water-Mother watched its weary wingbeats. Its bright, beady eye met her own. And between them there leaped a flash of perfect, wordless sympathy: a silent communication of god with god.

In response, the giantess lifted one of her knees out of the water. The bird wheeled, called out a startling, lonely cry, then landed easily on the great, smooth mound. However large the teal and however wide its wingspan, it was dwarfed by the magnitude of the Water-Mother's kneecap. The creature stood still for a moment, then settled itself and began to construct a nest from its own soft feathers, as if satisfied that it had found the perfect place at last.

The Water-Mother floated, watching. First the bird produced a golden egg, smooth and shining. Then out slipped a second, as golden as the first and as round as the goddess's belly. Then came a third, a fourth, a fifth and sixth, all identical. Finally the teal brought forth a seventh egg which was different from its fellows, a dark and heavy egg of iron, as dull and gray as the others were bright and burnished. If there were any doubt that the bird came from a sphere outside the common realm of nature, this strange progeny would put it to rest.

The teal fluffed out its feathers and sat, broody, on the Water-Mother's knee, keeping its eggs warm and away from the chill of the waters. Any communications that might have passed between them, bird and giantess, were unrecorded, but they bore each other company in that period of tranquil waiting.

Then, gradually, the time of peace came to an end. The clutch of eggs had at first been as cool as the metals they were made of. But as the bird sat over them the eggs grew warm. Then they gradually became hotter and hotter. Soon they glowed red, then white-hot, burning the smooth flesh of the Water-Mother's knee. Goddess she might be, but she was not immune to pain. With a roar that ruptured the air, she lashed out violently in agony. Her frenzied kicking churned up the waters into great tidal waves and one by one the seven burning eggs slipped from the nest and rolled off her knee. When they hit the surface of the ocean, they smashed and shattered.

Stricken, the Water-Mother looked about her for the teal. But the mysterious bird had flown.

Then something wonderful happened. The shattered fragments began to rise from the waters. And not a scrap of shell, nor a drop of the contents, were lost. From half the opalescent golden shells came the earth itself, from the rest came the cap of the heavens, curving over it. Bright yellow yolk shone out as the sun, the whites formed the moon, and whatever specks and spots remained rose up to become the stars and the clouds. The iron egg's dark remnants were transformed into the storms that blackened the sky.

Once again the Water-Mother drifted, ponderous and gravid. But now she had the sun to warm her and the moon to light her nights. Serene, she began to dream again. Countless times she saw the sun come up, countless times she saw the moon supplant it. The unborn child remained, as ever, within her.

Warmed by the solar orb, pulled one way and another by the moon-driven tides, the Water-Mother felt her time approaching. She followed the example of the teal and began to prepare a nest for her own offspring. She lifted up her head and looked about her. She reached out her massive hand and wherever her all-powerful fingers pointed there rose up jutting headlands, cliffs, coasts and promontories; the boundless sea was at last embraced by a shore.

She spread out her palm and shaped the sandy beaches, the dunes behind them, the flatlands and the gently rising downs, the higher hills and the towering mountains. With her fingernails she traced patterns on the faces of the mountains, carving out clefts and fissures and striations in the rock.

When she had finished this work, the giant goddess turned head over heels and plunged down under the surface of the sea with an almighty splash that nearly soaked the brand-new heavens. Swimming underwater, diving deep, she made the ocean floor, opening undersea caves

Waiting to bear the child begotten upon her by the wind and the waves together, the
Water-Mother floated in a shoreless sea. For centuries she drifted alone until a teal from a place
unknown flew out of the sky looking for a nesting-place.

and grottoes that would serve as hiding-holes for little fish and lurking-places for the bigger fish that would come to hunt them. She planted islands, both large and small, as well as rocks rising out of the water and hidden reefs that would some-day be the bané of mariners.

Finally she drew up stones from the sea bed and whirled and shaped them in her powerful hands to make the four massive pillars that would forever hold the sky above the land. Then she paused and looked about her, and smiled as if she liked what she saw. Her work of creation was complete.

Suddenly she felt violent stirrings in her womb. The child she had car-ried so long was growing impatient. Its own time had come to be part of the world its mother had made. She groaned as she felt it kicking and pushing deep in-side her, battering against her bones as if they were the bars of a prison. Gasping, she arched her body as her child thrust its way toward freedom and light.

Everywhere was blood and foam and a great churning of the waters, then all was blackness as the goddess ended her birth-agonies. Now the Water-Mother looked down into the sea and saw her newborn son, Vainamöinen. So long had he tarried inside her that he had grown to full size—indeed he had grown old within her womb. From the moment of his birth he possessed wisdom, a well-lined forehead and a long, venerable white beard.

No one knew for certain what the Water-Mother said to her son by way of welcome. But it was just as well that he came to her fully grown, needing no nursing, no swaddling, no tending, no teaching. Because there was plenty of work to be done. The Water-Mother had made the world, but she left it to her son to clothe and cover it.

And so the goddess floated upon the waters for a well-earned rest, watching her son Vainamöinen as he flexed his muscles, taxed his wits and summoned up his powers. She saw him bring forth grains and grasses, meadows of hay, fruits and flowering trees. He filled the sea with creatures and populated the land. He put pine trees on the hilly slopes, heather among the rocks, cherries where the earth was damp, bright-berried junipers on stony ground, rowans in the holy places, willows on the fens, tall oaks along the riverbanks, and all the living green things that would make food and fuel and shelter in the places where people would live. And he, Vainamöinen, son of the Water-Mother, would be the first among them: the first farmer, the first forester, the first gardener and, above all, the first poet and storyteller to hand down the memory of the beginning of the world for all time ☆

A Terrestrial Paradise

The massive Water-Mother, floating in her primal sea, had many counterparts in Creation lore. The Chinese too spoke of a giant, but theirs was of the opposite sex and the void he moved in was not water but a swirling cloud of chaos.

That the dwellers in the lands on the banks of the Yellow River should picture nothingness in this way was fitting. Year after year, billows of ocher-colored dust, tall as ten pagodas, blew into northern China from the high Mongolian steppes. When the dust clouds came, plows were abandoned in the fields.

Peasants, muffled to the eyes in dark cotton scarves, urged their plodding oxen back along the rutted lanes, anxious to reach the safety of their houses before the choking storm engulfed them. They sealed their doors and windows tight against the dust that erased the earth and sky and, over time, changed the face of the landscape itself.

This, they believed, was what the whole world had looked like in the very beginning, when the giant P'an Ku appeared out of nowhere. He had no known parentage but simply took shape, slowly, inside a whirlpool of dust and confusion. At first he was fast asleep. But the clouds of impenetrable matter must have nourished him, for while he was sleeping he started to grow. His head became a massive globe, his arms and legs expanded both in length and thickness to unim-

aginable dimensions. Then something must have woken him, perhaps one of his own enormous snores, deafening as a thunderclap. His huge eyes blinked open. All he could see before him was darkness and disorder. In a fit of pique he lifted his mountain of a fist and smashed the murk into countless tiny pieces.

Instantly the clouds cleared. The shattered fragments of chaos floated gently apart. The pieces that were *yang*—light, bright and hot—flew upward and became the sky. Those that were *yin*—hard, dark, cold and heavy—dropped down to form the earth. And P'an Ku drew himself up and stood between them. His feet were planted on the ground, his head supported the dome of the heavens. As he stood there—and he stood for a long time; some said as long as eighteen thousand years—the sky rose up higher and higher and the earth became thicker and heavier. Just as he had grown inexplicably bigger in his sleep, P'an Ku, when wide awake, continued to ascend even higher. Eventually he formed a living column, thousands of leagues tall, that kept the roof of the world from crashing downward and crushing the earth.

To pass the time, P'an Ku made a mallet and chisel and used them to chip away at the universe, carving it into its neat and proper shape. He may not have been alone while he worked, for certain old tales spoke of three magical beasts emerging from some unexplained elsewhere to bear him company: a dragon, a phoenix and a tortoise, that for all

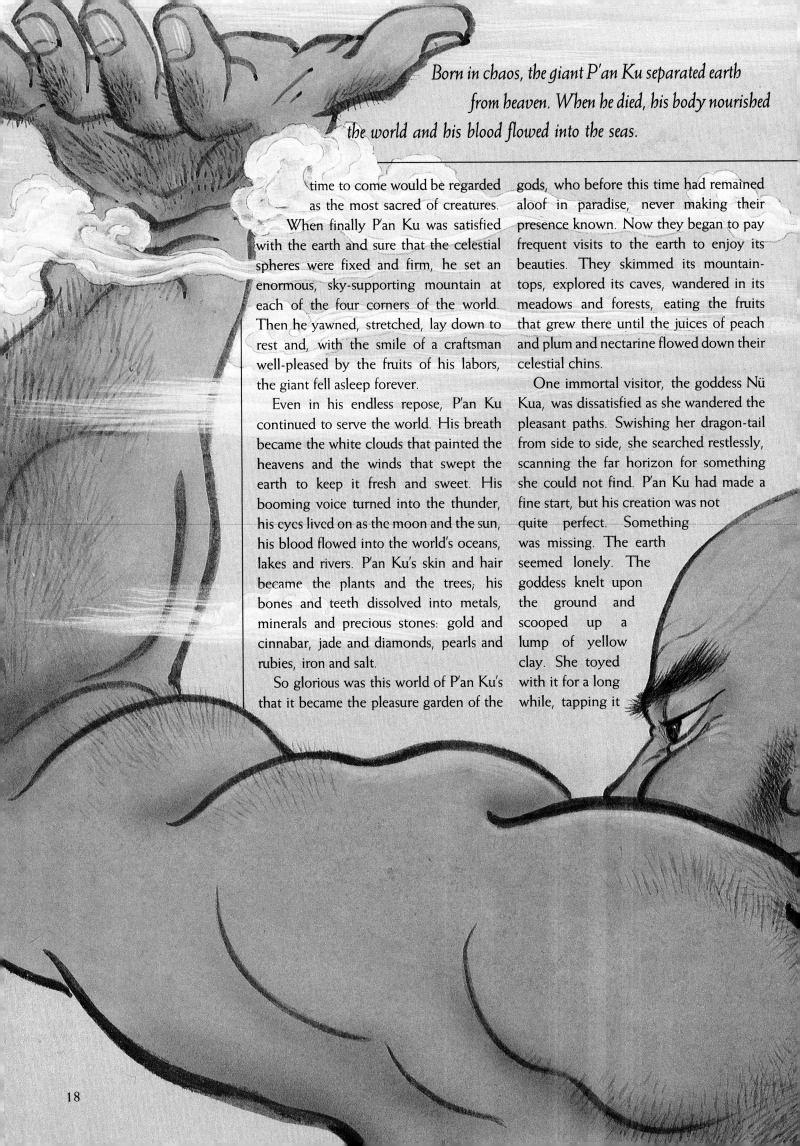

Born in chaos, the giant P'an Ku separated earth from heaven. When he died, his body nourished the world and his blood flowed into the seas.

time to come would be regarded as the most sacred of creatures.

When finally P'an Ku was satisfied with the earth and sure that the celestial spheres were fixed and firm, he set an enormous, sky-supporting mountain at each of the four corners of the world. Then he yawned, stretched, lay down to rest and, with the smile of a craftsman well-pleased by the fruits of his labors, the giant fell asleep forever.

Even in his endless repose, P'an Ku continued to serve the world. His breath became the white clouds that painted the heavens and the winds that swept the earth to keep it fresh and sweet. His booming voice turned into the thunder, his eyes lived on as the moon and the sun, his blood flowed into the world's oceans, lakes and rivers. P'an Ku's skin and hair became the plants and the trees; his bones and teeth dissolved into metals, minerals and precious stones: gold and cinnabar, jade and diamonds, pearls and rubies, iron and salt.

So glorious was this world of P'an Ku's that it became the pleasure garden of the gods, who before this time had remained aloof in paradise, never making their presence known. Now they began to pay frequent visits to the earth to enjoy its beauties. They skimmed its mountain-tops, explored its caves, wandered in its meadows and forests, eating the fruits that grew there until the juices of peach and plum and nectarine flowed down their celestial chins.

One immortal visitor, the goddess Nü Kua, was dissatisfied as she wandered the pleasant paths. Swishing her dragon-tail from side to side, she searched restlessly, scanning the far horizon for something she could not find. P'an Ku had made a fine start, but his creation was not quite perfect. Something was missing. The earth seemed lonely. The goddess knelt upon the ground and scooped up a lump of yellow clay. She toyed with it for a long while, tapping it

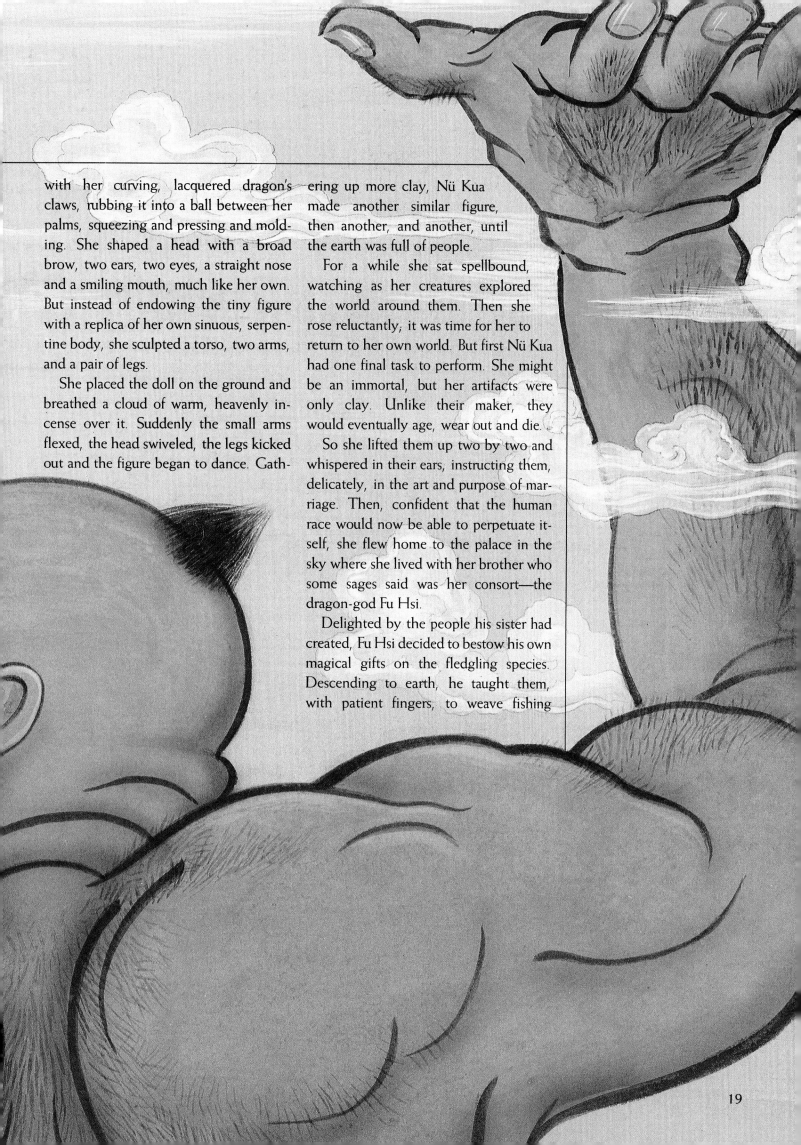

with her curving, lacquered dragon's claws, rubbing it into a ball between her palms, squeezing and pressing and molding. She shaped a head with a broad brow, two ears, two eyes, a straight nose and a smiling mouth, much like her own. But instead of endowing the tiny figure with a replica of her own sinuous, serpentine body, she sculpted a torso, two arms, and a pair of legs.

She placed the doll on the ground and breathed a cloud of warm, heavenly incense over it. Suddenly the small arms flexed, the head swiveled, the legs kicked out and the figure began to dance. Gathering up more clay, Nü Kua made another similar figure, then another, and another, until the earth was full of people.

For a while she sat spellbound, watching as her creatures explored the world around them. Then she rose reluctantly; it was time for her to return to her own world. But first Nü Kua had one final task to perform. She might be an immortal, but her artifacts were only clay. Unlike their maker, they would eventually age, wear out and die.

So she lifted them up two by two and whispered in their ears, instructing them, delicately, in the art and purpose of marriage. Then, confident that the human race would now be able to perpetuate itself, she flew home to the palace in the sky where she lived with her brother who some sages said was her consort—the dragon-god Fu Hsi.

Delighted by the people his sister had created, Fu Hsi decided to bestow his own magical gifts on the fledgling species. Descending to earth, he taught them, with patient fingers, to weave fishing

The dragon-tailed goddess Nü Kua gave earth
 its first human inhabitants. She sculpted their bodies from clay, infused
 them with life and taught them how to love.

The god Fu Hsi imparted skills to the new worldlings that would enable them to survive and prosper. He taught them how to make nets for fishing, invented an alphabet for their use and instructed them in the art of writing.

nets so they might feed themselves from the sea. To please them in their idle hours, he devised the lute; by plucking on its strings they could reproduce for themselves the sweet sounds that wafted through the courts of heaven. And, lest they suffer from cold, he showed them how to make fire by drilling wood against wood. Last of all, he taught the people the art of writing, so that they could record their knowledge for those who would come after them.

But, as so many histories of the world's beginning related, this new earthly paradise was nearly destroyed in a holocaust. The Chinese storytellers attributed the cause of the cataclysm to a war between two rival cosmic powers—the spirit of fire and the spirit of water.

The conflict began in heaven but spread swiftly to earth. When it ended, the goddess Nü Kua descended to view the devastation at close quarters. She found her beloved world in ruins. Its once smooth surface was scored with fissures, scarred with deep cracks and canyons. The forests were on fire. The choking dust from which the world had first emerged now appeared to be swallowing it up again. Floodwaters menaced the lowlands. Wild animals roamed everywhere, devouring the frightened humans they encountered.

The whole earth tilted crazily. The sky itself was torn and tattered. Violent storm winds, howling out of heaven, rushed through the holes and buffeted the earth below.

The goddess immediately began to put the world to rights. Using her celestial powers, she doused the smoldering forests and piled up the ashes of burned plants and reeds to dam the floodwaters. She drove the wild beasts back to their lairs and soothed the terrified humans.

Then she turned to the more difficult task of repairing the badly damaged sky. She gathered up smooth, shining stones, choosing only those that were red, yellow, blue, white or black, which from then on would be the primary colors of the painter's palette. Some of these stones she ground to a fine powder in a giant mortar and then mixed the powder into a plaster to mend the narrowest cracks. Others she placed in a vast caldron, melted them down and refined them into a metal that glowed like jade but was harder than iron. With those same deft dragon-claws that had fashioned humankind, the goddess carefully patted the substance into place to patch the gaping holes in heaven.

Now the world and its creatures would be safe until the end of time. Ever after, on summer nights when the sunset was most glorious, her children still looked at the sky's glowing colors and recognized her handiwork. Finally, most important of all, she swept away the swirling clouds of dust and debris, and the world again emerged from the tide of chaos ☆

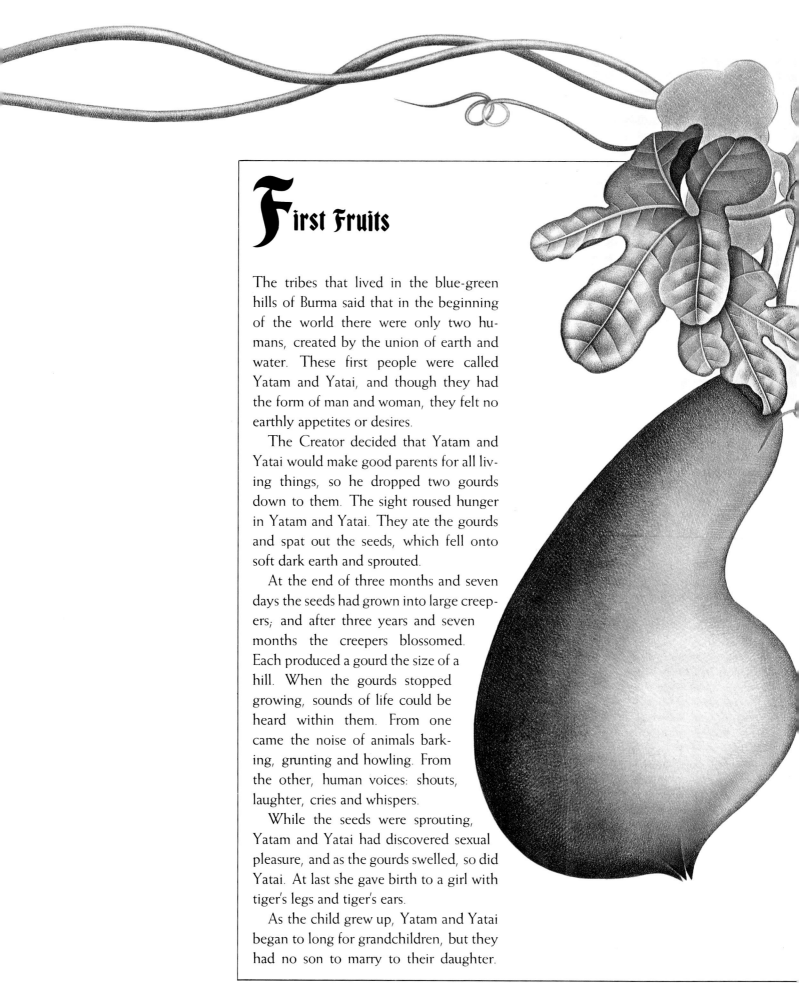

First Fruits

The tribes that lived in the blue-green hills of Burma said that in the beginning of the world there were only two humans, created by the union of earth and water. These first people were called Yatam and Yatai, and though they had the form of man and woman, they felt no earthly appetites or desires.

The Creator decided that Yatam and Yatai would make good parents for all living things, so he dropped two gourds down to them. The sight roused hunger in Yatam and Yatai. They ate the gourds and spat out the seeds, which fell onto soft dark earth and sprouted.

At the end of three months and seven days the seeds had grown into large creepers; and after three years and seven months the creepers blossomed. Each produced a gourd the size of a hill. When the gourds stopped growing, sounds of life could be heard within them. From one came the noise of animals barking, grunting and howling. From the other, human voices: shouts, laughter, cries and whispers.

While the seeds were sprouting, Yatam and Yatai had discovered sexual pleasure, and as the gourds swelled, so did Yatai. At last she gave birth to a girl with tiger's legs and tiger's ears.

As the child grew up, Yatam and Yatai began to long for grandchildren, but they had no son to marry to their daughter.

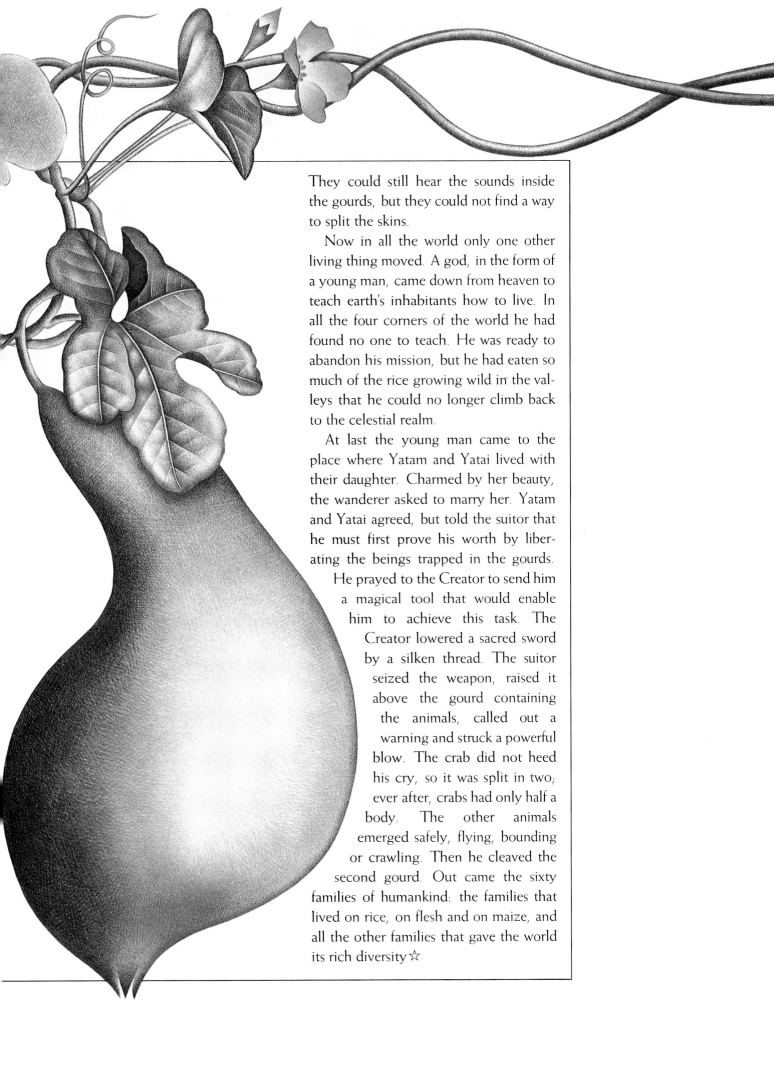

They could still hear the sounds inside the gourds, but they could not find a way to split the skins.

Now in all the world only one other living thing moved. A god, in the form of a young man, came down from heaven to teach earth's inhabitants how to live. In all the four corners of the world he had found no one to teach. He was ready to abandon his mission, but he had eaten so much of the rice growing wild in the valleys that he could no longer climb back to the celestial realm.

At last the young man came to the place where Yatam and Yatai lived with their daughter. Charmed by her beauty, the wanderer asked to marry her. Yatam and Yatai agreed, but told the suitor that he must first prove his worth by liberating the beings trapped in the gourds.

He prayed to the Creator to send him a magical tool that would enable him to achieve this task. The Creator lowered a sacred sword by a silken thread. The suitor seized the weapon, raised it above the gourd containing the animals, called out a warning and struck a powerful blow. The crab did not heed his cry, so it was split in two; ever after, crabs had only half a body. The other animals emerged safely, flying, bounding or crawling. Then he cleaved the second gourd. Out came the sixty families of humankind: the families that lived on rice, on flesh and on maize, and all the other families that gave the world its rich diversity ☆

Fall of the Sky Maiden

The movements of stories across the world's oceans were as mysterious as the migrations of birds or the spawning journeys of the eel and the salmon. However alien the tongues they were spoken in, however far apart the places they were told, many tales of the world's beginnings echoed one another. It was as if the same seed sprouted everywhere into flowers whose variations in shape and color reflected the different climates that had nourished them.

In so many tales, similar figures appeared: The Water-Mother of the Finns, the sibling dragon-gods of China, and Yatam and Yatai's sword-wielding son-in-law, miraculous beings who served not as the world's makers but as its midwives, were all said to have descended from the heavens. And far away across unmapped oceans, the chroniclers of the Iroquois nation identified the catalyst of Creation as a woman who fell—or perhaps was pushed—from a homeland in the sky.

Unlike the other beings of Creation lore, she did not come from some nebulous, otherworldly paradise. The land of her birth was nearly as familiar to the Iroquois as their own, although it hovered somewhere far above them. It was a place whose inhabitants lived much as the Iroquois did. They hunted animals, wore skins for warmth, relied on chief-tains and medicine men to guide them. But if it had not been for one of their princesses, Ataentsic, daughter of the Sky Chief, theirs would have been the only world and ours would never have existed.

The story began in that upper world when Ataentsic was stricken by a strange and terrible illness. Its symptoms were mysterious: The young woman lay for a long time, unable to move or speak, her eyes unblinking, her pulse rate so slow as to be almost nonexistent, her breathing shallow. No one knew what spirit she had offended or what curse had claimed her, so she was left to languish alone in her tent, isolated from the rest of the tribe, lest the contagion spread.

She made no sign of recognition when the doeskin flap of her tent opened to admit her father and the medicine man. The fresh air of spring drew in the scents of flowers and the call of birds, but Ataentsic barely remembered what they signified. She felt dry and ancient, void of motion and energy, but she lived on.

The people who were gathered around her pallet began to talk. Her father's voice boomed deep and thunderous. "Rise up, my daughter, and live again with us," he commanded.

Where once she could not fail to heed his merest whisper, now even his thunder

In the land above the heavens, a woman of royal blood sickened with an illness that feigned death itself. She was placed in a grave within the roots of a sacred tree, and her departure from one world signaled her entrance into another.

was nothing to her. The Sky Chief growled fiercely to the wise man, "Why will she not obey?"

The shaman's face, its wrinkles forming the same pattern as the carvings on the stick he held in his hand, waited before he spoke. He shook his beaded medicine pouch over her face and a sifting of fine dust fell into her eyes, but she did not blink. Finally, shutting his eyes against the power of Ataentsic's empty stare, the shaman entered the world of visions.

The Sky Chief waited, humble in the presence of powers greater than his own. After a time, the medicine man emerged from his trance and spoke the message that the spirits gave him.

"She lives but she is dead. And we cannot keep the dead among us. Beneath the sacred tree at the center of our world we must dig a grave. Deep in the roots of the great tree we will bury her."

"So be it," intoned the Sky Chief.

If he felt grief at the loss of his daughter, or fear of the damage that her burial might do to the roots of the sacred tree, neither his face nor his voice revealed it. The fate of the entire tribe depended on unquestioning obedience to the commands received by the shaman from the spirits.

But the people's well-being also depended on the tree itself. Its fruit nourished them, its shade protected them. To tamper with the tree might destroy them all. Nevertheless, the shaman conveyed the message of the spirits and, with much labor, the tree's roots were exposed and the grave prepared.

Keening and moaning, the princess's tribe gathered outside her tent, mourning her death. Inside the tent, the women dressed her seemingly lifeless body in the softest and most beautiful furs. Then they shouldered the pallet she lay on and carried her out of the tent and into the glare of the sun. Ataentsic's eyes, locked open by her trance, were seared with brightness. The waiting crowd screamed and howled, trying to wake her from the sleep of the dead even as they carried her to burial. Silence and shadow surrounded Ataentsic as she was placed in the grave in a hollow within the tree roots

When the woman fell out of the land above the sky and our world was
nothing but air and water, the birds, that were its chief inhabitants,
came to her rescue and caught her in their wings.

and covered over with fragrant grasses.

Then, in quick succession, two fateful happenings occurred. With a creaking and rending sound, the great tree began to die. The roots shrank up and curled in upon themselves like enormous claws dragging at Ataentsic's flesh, while the ground trembled beneath her. And a moment later, a hand reached down into the open grave, penetrating the strewn grasses, and pushed Ataentsic violently into the earth.

Some versions of the tale related that the hand belonged to one of Ataentsic's previous lovers who had been rejected. Others suggested that an angry spirit guardian had come to punish the princess for her violation of the sacred tree. But whatever the agency, Ataentsic and the tree that held her in its roots fell down together through a deep hole in the floor of the upper world.

Just as she passed through the aperture, Ataentsic's illness lifted and her senses returned. Legs, arms and fingers regained their mobility, but she was powerless to help herself in her fall. She reached out wildly, catching scraps of earth, stones, roots and seed. She felt the soft bodies of burrowing animals lodge in her hair and her clothes. Finally, there was nothing more to catch but air.

So began her long journey. Ataentsic, clinging to the tree, fell through clouds and past rainbows until a vast sea appeared beneath her. Birds flew through the void, the only inhabitants of this unknown world. A great swan, its long neck outstretched, circled her. A tern landed for a moment on the tree and looked into her eyes, unblinking.

Ataentsic screamed as the rushing winds tore the tree from her grasp and sent it crashing down into the water. Ataentsic continued falling alone.

But according to the lore of the Iroquois, humans were not the only creatures endowed with wisdom. The circling birds, in silent agreement, drew closer together to form a net of beating wings and caught Ataentsic as she fell.

At that moment a giant turtle rose from the sea, huge as an island. Perhaps it was simply disturbed in its sleep by the great tree's fall, or perhaps it was sent by some benevolent spirit. But with the generosity of a king secure in his own realm, the turtle offered its back as a landing place. The birds set Ataentsic gently down on its shell.

Then the creatures from the upper world that had hidden in Ataentsic's hair and clothing and fallen with her emerged from their hiding places. The birds preened their mauled and matted feathers and the creatures of the water—the otter, muskrat and beaver—pulled themselves up onto the turtle's back.

But the shell was too curved and slippery to provide a foothold for the many animals that settled upon it. So, in the wordless language of the ancient beasts, the turtle commanded those animals that could swim to dive down to the place where the great tree from the upper world now rested under the water. It bade them bring up some of the earth that still clung to its roots.

Perched precariously on the back of a giant turtle, the
Sky Woman and the creatures that had descended with her
floated in an endless sea, unsure of survival.

The otter was the first to try, but the sea was as deep as it was vast and nothing more was seen of it until it came to the surface, dying of exhaustion. It had never reached the bottom at all. The beaver tried next, but it too failed, and died beside the otter. The muskrat resolved to try, but soon its furry body lay next to the others. None of them had reached the tree. There was no earth to cover the shell. Ataentsic clung fast, but the beasts around her stirred restlessly.

Then the frog, even though it was smaller than the others, dived into the sea. It disappeared in the watery depths and Ataentsic waited, shivering in the cold gray dawn. When the surface of the waters broke at last, the frog appeared, barely alive. But as it took its dying breath, all those creatures on the turtle's back could see a speck of earth from the great tree clinging to its mouth.

The tern flew down and took the dirt gently from the dead frog's jaws and placed it on the turtle's shell. Ataentsic stood and walked in a circle around the lump of earth, gathering all the upper-world magic she could muster and forcing it into her footsteps. As she paced, the lump of dirt expanded in all directions until the turtle's back was entirely covered and the surface of the earth was formed.

The great turtle closed its eyes to sleep and float for all eternity, returning to the somnolence of perpetual hibernation. Soon the earth on its back began to sprout with the plants whose seed had fallen down with Ataentsic from the upper world.

But these were not the only seed that germinated in this new place. There was human seed as well, growing inside the womb of Ataentsic.

Each tribe and each storyteller in the Iroquois nation gave a different version of how the Sky Woman came to be with

child. Some said that she carried her burden down with her from the sky. However it happened, there was not one but two children inside her, waiting to be born. And even in the womb they were already locked in the conflict that was to rule their lives and shape the earth.

It was said that children born to the people of the upper world had the gifts of sense and speech from the moment of conception. Ataentsic's offspring, deep within their mother, saw faint traces of daylight and decided that it was time to emerge from their gestation. They began to swim toward the source of the light, but they soon realized they were heading in different directions. One brother said that he saw the light coming from above and wanted to be born from his mother's mouth or nose. His twin warned him that the birth canal lay in the opposite direction. To be born any other way would kill their mother.

But the stubborn sibling did not listen. He forced a passage out from under his mother's arm. As his twin had predicted, his unnatural birth killed Ataentsic. The second brother emerged in the proper manner, then buried and mourned the mother he had never known. From that day, the brothers hated one another.

Here too, as in the tale of the Water-Mother, the work of completing Creation was passed to a second generation. The two sons of Ataentsic began to shape the features of the earth.

Loathing one another, yet inextricably bound together, the siblings—whom the storytellers named Sapling and Flint— entered into a duel. Sapling embodied the force of creation, Flint of destruction, and their conflict gave the world its form.

Sapling first made brooks and rivers. Then he created the hills and mountains to provide a path for these waterways that would allow them to run straight down to the sea. Flint could not stand such efficiency and order, so he scattered and splintered the mountains and bent the course of the rivers, tossing boulders in their paths to create rapids and waterfalls.

Sapling next covered the hills with forests and the plains with fruit trees. This beauty infuriated Flint, so he gnarled the trees and hung them with briars. Then he dived into the ocean and filled it with sea monsters and storms.

Sapling replied by creating eagles and grazing animals to eat the grass on the plains. Flint made mountain lions, foxes, bears and wolves to harass them. As an afterthought, he created vultures and crows, so that evil could fly as well.

Some chroniclers related that Sapling killed his wicked brother by striking him with a stag's antler. Others insisted that Flint was wounded but not killed, for evil could never die. But all the tale-tellers agreed that their fight was necessary.

Without hunters and predators, the grass-eaters would ravage the land and finally starve. Without the plants that made poison, there could be no healing drugs or antidotes. And without sea monsters, storms and dangerous river-journeys, there would be no adventures, no need to triumph over obstacles and, worst of all, no work for storytellers ☆

The Trembling Earth

Whenever the ground quaked and trembled, demolishing houses, uprooting trees and causing great abysses to open, it served as a reminder of the earth's precarious position, for ancient lore made it plain that the world did not rest on firm foundations.

Some African storytellers said that all life sprouted from the head of a living giant. Trees, flowers and grasses grew as his hair, and people and animals were the parasites crawling on his scalp. Earthquakes occurred whenever the giant sneezed or twitched or turned his head too suddenly. Elsewhere it was thought that a foul-tempered giant bore the earth on his back and sometimes fell to fighting with his equally enormous brothers—who themselves may have carried planets on their backs. Yet some wise elders insisted that the giant was more affectionate in nature; when he and his wife embraced, they did so with such vigor that the whole earth quaked with their passion.

A Hog's Itch

Perched precariously in a vulnerable place on the earth's unsteady crust, the South Sea Islanders looked to their storytellers for reassurance when the mountains cracked and the earth swayed under them. In the Celebes, they believed quakes occurred because the world rested on the bristly back of a great hog. When an itch spread down its spine, it rubbed its back on a rough-barked palm tree growing in a cosmic grove that was far too massive for ordinary mortals to comprehend. The world heaved and rolled. Any subterranean rumbling that accompanied the tremors was the sound of the hog itself, grunting with satisfaction that its itch had at last been scratched.

A Restless Frog

In Mongolia, wise elders had their own explanation for tremors that disturbed the broad expanses of their homeland and rattled the frames of the round felt yurts they lived in. They described a great frog who balanced the world on its back. Sometimes a shiver ran under its smooth green skin, and its muscles, longing for a jump, twitched and trembled. Slipping from one side of its slimy back to the other, the earth quaked.

It was fervently hoped by all mortals that the creature would control its restless impulses and never take it into its head to leap into the air or dive into the waters of a pond. If it ever did, the results would be too catastrophic to contemplate.

Fiery Lairs of Gods and Monsters

Wherever mountaintops breathed fire, tales were told about the terrifying and random violence of volcanoes. The long-memoried Greeks knew how destructive they could be. Back in the mists of distant time, a whole island civilization, Atlantis, had vanished overnight in a cataclysmic explosion. And nearby, on islands and coasts throughout the Mediterranean, there were other mountains that now and again exhaled smoke and thunder, and spat out molten rock. Who knew what monstrosities lurked beneath, caged by the weight of the rock, struggling to break their bonds?

Fortunately, not all volcanoes were the abode of violent and vengeful spirits. Vulcano itself, a hill of ash and rubble on an island north of Sicily, was the home of Vulcan, the Roman god of fire and the blacksmith to all the deities. Here the bearded god, with his hammer, tongs and anvil, forged the breastplate of Hercules, the shield of Achilles and the arrows of Apollo and Artemis. To the Romans, Vulcan was a beneficent deity, well able to control the fires he used. Seldom did he allow the furnace to spew out molten rock, and Sicilians became accustomed to the deep and distant rumbles as he pursued his ancient trade.

Most volcanoes, however, contained more fearsome tenants. Sicilians knew only too well the destructive powers of Mount Etna. They blamed its violence on a terrible monster named Typhon whose limbs were serpents, whose hundred heads were those of dragons, whose wings blotted out the sun, and whose offspring were the winds. His name long survived him in the greatest wind of all, the typhoon.

No one knew for sure how the creature came into being. Some claimed Typhon was the son of Gaea, the earth-goddess from whom sprang a brood of monsters, the Titans. Others said that there were two such creatures, father and son.

In any event, Typhon's powers were so awesome that he challenged the established divine order, opposing Zeus himself. While the lesser Greek gods cowered out of range, the two giants fought. Zeus attacked first, wielding a sickle of adamantine strength and sharpness. Typhon wrested it from him and succeeded in cutting out the sinews of Zeus's hands and feet. Typhon then bound the crippled Zeus and shut him away in a cave, with his sinews wrapped in a bearskin and placed tantalizingly close to where the prisoner lay, guarded by a fearsome dragon.

There, Zeus's son Hermes, always the willing helper and the bearer of good fortune, found him. Hermes, youthful, athletic and strong, defeated the dragon and returned his father's sinews to make him whole again.

Zeus, burning for revenge, pursued Typhon to Sicily, caught him, bound him and buried him deep in the underworld beneath Mount Etna.

Forever after, Typhon sought to escape, writhing and beating at the gates of Hades, occasionally belching fire and brimstone over the vineyards and fields surrounding the mountain ☆

Defeated in a cosmic battle, the monster Typhon was imprisoned in Sicily's Mount Etna. Whenever he grew restless, the mountain rumbled and sent out streams of molten lava.

Phaethon's Folly

The sun's daily procession through the sky was no mystery to the dwellers in the temperate lands around the Aegean's rim. They believed that every morning a radiant god mounted a fiery chariot and guided a team of spirited horses—themselves divine—across the arc of the heavens, from east to west.

Unlike the inhabitants of the chilly wildernesses to the north, these people had no fear that their sun-king would ever disappoint them or die in combat with the powers of cold and darkness. The appearance and transit of his golden chariot were as predictable as the steps of a scrupulously executed ritual dance.

But a time came when the inhabitants of the warm lands were shaken out of their complacency. The solar god they called Helios was forced to deviate from his sacred routine. He did so only once, and by no choice of his own. But when he relinquished the control of the sun-bearing steeds to another charioteer, the consequences were catastrophic. The fiery cataclysm that ensued was seared into the memory of many of the world's nations.

The terrible chain of events, said an early poet, began when a young stranger arrived at the entrance to the glorious mountaintop palace of the sun. He paused for a moment, panting after his climb, to contemplate the massive edifice that stood before him. Along its lofty porches, columns of gold and bronze rose up to meet graceful pediments and sloping ivory roofs. Then he approached the tall, elaborately carved doors, squared his shoulders, knocked once and pushed the portals open.

The sight that greeted him was so brilliant as to be unbearable. He shielded his eyes from the blinding glare. At the far end of the hall, the sun-god sat in splendor on an emerald throne. His servants, the spirits who governed the hours, the days, the months, the years and the generations, were in attendance. His four chief ministers, the seasons, exchanged curious glances at the sight of the young intruder.

The sun-god asked the stranger his name and business. The boy hesitated. Whatever courage had brought him to this place momentarily deserted him. Then he gathered his wits, took a deep breath and told his story. His name was Phaethon—the Shining One. He had never known his father, but his mother claimed that he was the child of the sun-god himself. She had made no secret of that brief liaison: To lie with a god was a mark of honor Phaethon, with the arrogance of youth, had boasted to his schoolmates of his semidivine origins. They jeered and dismissed him as a fatherless braggart with a liar for a mother. Stung by their gibes, he left the place of his birth and made his way to Helios' palace to find out the truth for himself.

Helios commanded the boy to come closer, looked him up and down, then enveloped him in a fiery, fatherly embrace. There could be no doubt about it, the god asserted. Phaethon was indeed his son. And as a token of paternal affection, he would grant his child anything he wished. Let no one think the gods of Olympus forgot their children.

Phaethon smiled and pondered for a moment. He admitted that he did have one request. The sun-god swore upon the terrible lake in the Nether World—that infernal pool by which all immortals took their oaths—that he would do whatever Phaethon asked. The boy drew breath and spoke. He wished to drive the sun-chariot across the sky.

Helios paled. Only he alone among gods had the power to control the blazing vehicle on its perilous journey. No force in the world was as powerful or as dangerous as the sun. Even he himself, he confessed to Phaethon, was awed and frightened every time he drove the chariot. Helios begged his son to choose another favor. But the boy was adamant. He knew that a god's oath could never be broken.

He reminded his father of the vow sworn on the dreaded lake, and asked what the consequences would be if he reneged on his promise. The sun-god did not answer him, but in his father's eyes Phaethon saw terror, anguish and a dark foreboding. Helios shook his head in shame and sorrow for his own hasty, ill-considered offer. Reluctantly, he led his son to the golden chariot and the four winged horses that were breathing fire and stamping with impatience to be off.

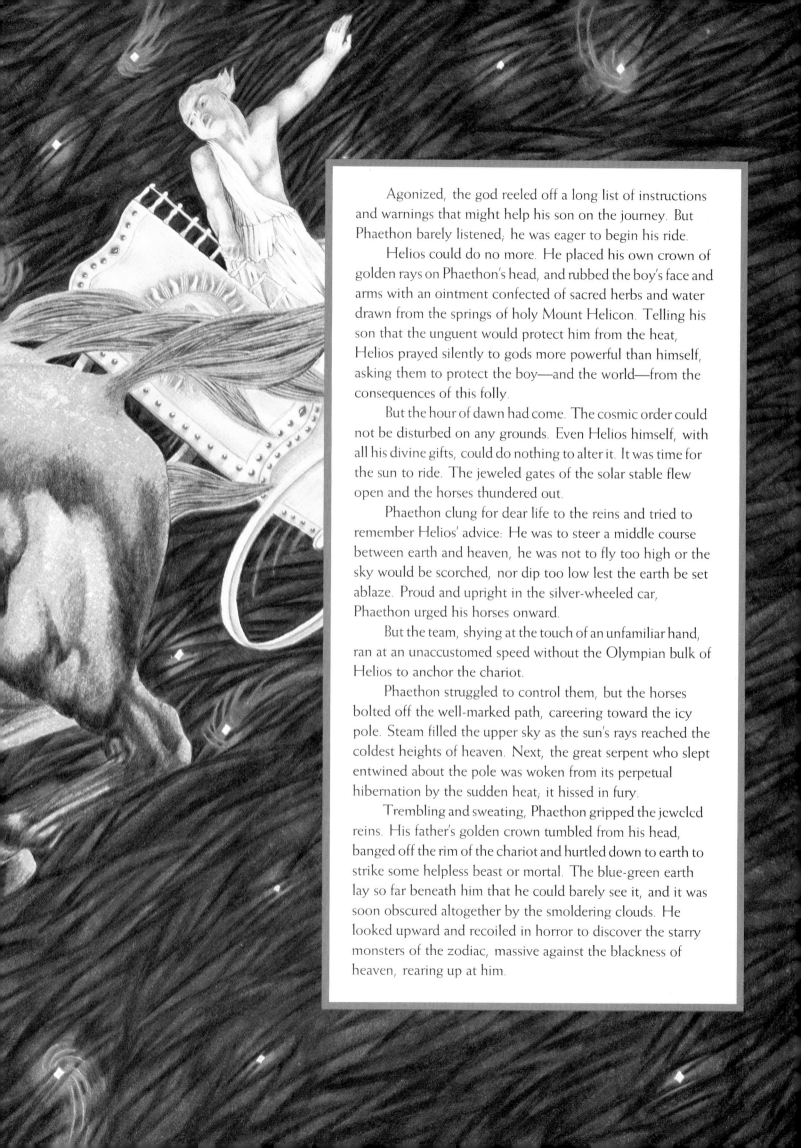

Agonized, the god reeled off a long list of instructions and warnings that might help his son on the journey. But Phaethon barely listened; he was eager to begin his ride.

Helios could do no more. He placed his own crown of golden rays on Phaethon's head, and rubbed the boy's face and arms with an ointment confected of sacred herbs and water drawn from the springs of holy Mount Helicon. Telling his son that the unguent would protect him from the heat, Helios prayed silently to gods more powerful than himself, asking them to protect the boy—and the world—from the consequences of this folly.

But the hour of dawn had come. The cosmic order could not be disturbed on any grounds. Even Helios himself, with all his divine gifts, could do nothing to alter it. It was time for the sun to ride. The jeweled gates of the solar stable flew open and the horses thundered out.

Phaethon clung for dear life to the reins and tried to remember Helios' advice: He was to steer a middle course between earth and heaven, he was not to fly too high or the sky would be scorched, nor dip too low lest the earth be set ablaze. Proud and upright in the silver-wheeled car, Phaethon urged his horses onward.

But the team, shying at the touch of an unfamiliar hand, ran at an unaccustomed speed without the Olympian bulk of Helios to anchor the chariot.

Phaethon struggled to control them, but the horses bolted off the well-marked path, careering toward the icy pole. Steam filled the upper sky as the sun's rays reached the coldest heights of heaven. Next, the great serpent who slept entwined about the pole was woken from its perpetual hibernation by the sudden heat; it hissed in fury.

Trembling and sweating, Phaethon gripped the jeweled reins. His father's golden crown tumbled from his head, banged off the rim of the chariot and hurtled down to earth to strike some helpless beast or mortal. The blue-green earth lay so far beneath him that he could barely see it, and it was soon obscured altogether by the smoldering clouds. He looked upward and recoiled in horror to discover the starry monsters of the zodiac, massive against the blackness of heaven, rearing up at him.

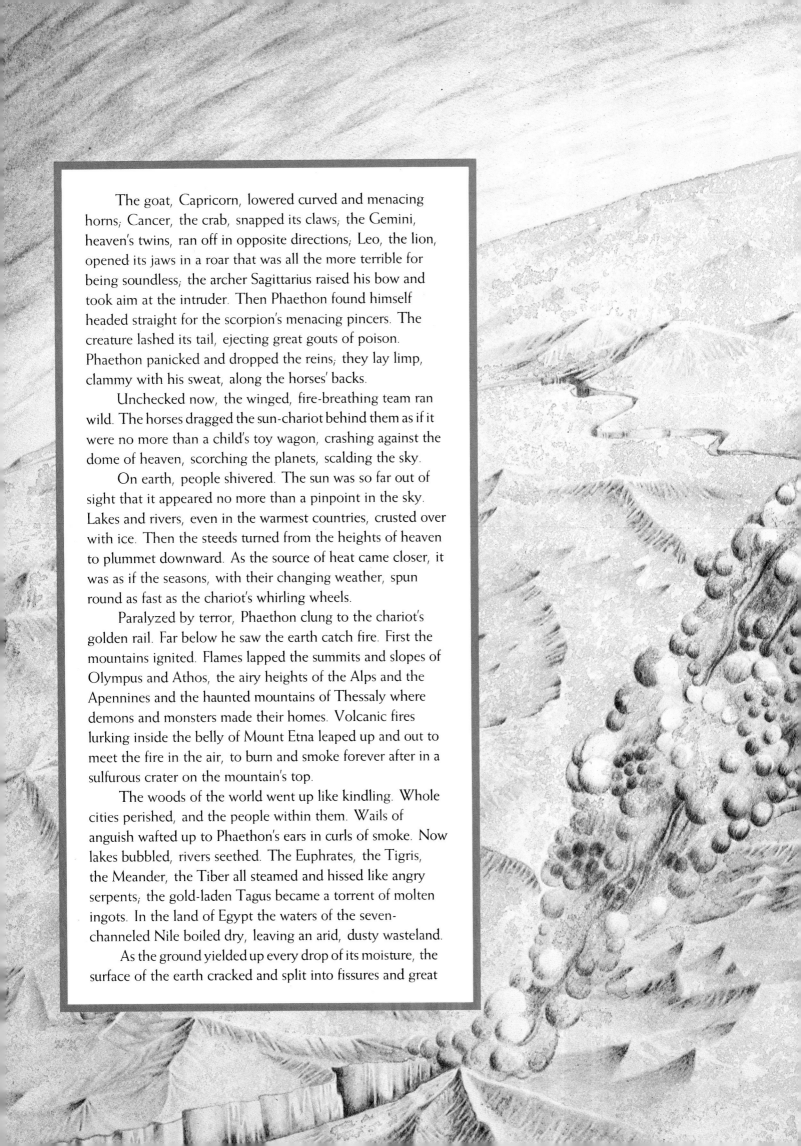

The goat, Capricorn, lowered curved and menacing horns; Cancer, the crab, snapped its claws; the Gemini, heaven's twins, ran off in opposite directions; Leo, the lion, opened its jaws in a roar that was all the more terrible for being soundless; the archer Sagittarius raised his bow and took aim at the intruder. Then Phaethon found himself headed straight for the scorpion's menacing pincers. The creature lashed its tail, ejecting great gouts of poison. Phaethon panicked and dropped the reins; they lay limp, clammy with his sweat, along the horses' backs.

Unchecked now, the winged, fire-breathing team ran wild. The horses dragged the sun-chariot behind them as if it were no more than a child's toy wagon, crashing against the dome of heaven, scorching the planets, scalding the sky.

On earth, people shivered. The sun was so far out of sight that it appeared no more than a pinpoint in the sky. Lakes and rivers, even in the warmest countries, crusted over with ice. Then the steeds turned from the heights of heaven to plummet downward. As the source of heat came closer, it was as if the seasons, with their changing weather, spun round as fast as the chariot's whirling wheels.

Paralyzed by terror, Phaethon clung to the chariot's golden rail. Far below he saw the earth catch fire. First the mountains ignited. Flames lapped the summits and slopes of Olympus and Athos, the airy heights of the Alps and the Apennines and the haunted mountains of Thessaly where demons and monsters made their homes. Volcanic fires lurking inside the belly of Mount Etna leaped up and out to meet the fire in the air, to burn and smoke forever after in a sulfurous crater on the mountain's top.

The woods of the world went up like kindling. Whole cities perished, and the people within them. Wails of anguish wafted up to Phaethon's ears in curls of smoke. Now lakes bubbled, rivers seethed. The Euphrates, the Tigris, the Meander, the Tiber all steamed and hissed like angry serpents; the gold-laden Tagus became a torrent of molten ingots. In the land of Egypt the waters of the seven-channeled Nile boiled dry, leaving an arid, dusty wasteland.

As the ground yielded up every drop of its moisture, the surface of the earth cracked and split into fissures and great

canyons, sending chinks of light into the darkest crevices of the underworld. On the verges of the purple sea, the gardens of the nymphs, their sacred springs and groves and fountains, were all burned to cinders.

Before his eyes, Phaethon beheld the once-green land of Libya blasted into an arid desert for all time. And, in the regions where the solar chariot swooped lowest, those people who did not perish outright were seared forever by the intense heat. Thus did the whole Ethiopian nation turn the color of polished jet and gleaming ebony. All this young Phaethon watched helplessly as his untrammeled horses whirled him through the air.

But this was not all that Phaethon saw. As he was dragged round by the team of panicking horses, he glimpsed, now and again, the awesome sight of the gods in anguish. Poseidon, father of the ocean, tried three times to raise his head above the waves and three times the burning air drove him back down again, to grieve for the dead seals and dolphins that now floated lifelessly on the water.

Gaea, the earth-goddess, groaned and shuddered. Her hair was scorched, her lips cracked and swollen, her eyes clogged with ashes. She pleaded with Zeus, king of the gods, to come to her aid. And Helios, the unwilling author of the catastrophe, stood by, powerless, tearing at his robes in his grief and mortification. When Phaethon flew past the sun-god he turned his head away, not having the courage to meet his father's eye.

What he met instead was the angry eye of Zeus. The divine monarch shook his crowned head in a gesture of regret, raised his arm and with unerring aim shot off a thunderbolt. And this thunderbolt flashing toward him was the last thing that Phaethon ever saw.

Pierced by a fork of celestial lightning, the mortal son of the deathless sun-god was flung from the chariot. He fell blazing from the sky, an evanescent shooting star.

The chariot finally shattered. The frightened steeds slipped from the splintered yoke, panting and foaming, then turned, exhausted, to find their own way back to the palace of the sun. Helios hid his head in shame and for a long time the world was in darkness☆

Two

Celestial Lore

Cleaving the Day

From the moment of Creation, the world was a dance of opposites: fire and water, heat and cold, earth and sky. The greatest of these polarities, the mother of all metaphors of difference, was the contrast between night and day.

To the Arabian poets, the day was a golden lion and the night a frightened roe deer that fled the great cat's claws as they ripped through the mists of morning. To the Magyar nomads who came riding out of Asia, night and day were creatures that were tethered like slaves or horses; each was unbound in turn to bring the twilight or the dawn. To the tow-headed seafarers of the Norse countries, who measured alternating periods of light and dark in months instead of hours, the night was a goddess who gave birth to day in long and painful parturition.

The two were not always separate entities. Some long-memoried chroniclers spoke of an age of constant daylight. Others told of a time when everything was black. In the deepest jungles of the south, where the temples and treasures of lost kingdoms lay concealed in forgotten caves or buried deep under a mesh of vines and creepers, storytellers recalled the dramas that brought brightness to a nocturnal world, or freed humankind from the glare of an endless morning.

Some tribes recalled a grim time of perpetual darkness, when all the light was jealously guarded by a family of birds that had to be bullied, cajoled and finally tricked into sharing day with their fellow creatures. In another part of the same jungle, people maintained that it was night, not day, that had to be requested; it lay not in the clutches of birds but within the coils of a giant anaconda, the Lord of Creation himself.

In the very beginning, those storytellers said, when the Great Serpent—the anaconda—created the forest and the rivers, there was no darkness anywhere except at the bottom of the deepest lakes. The sun remained fixed forever in the sky. It was always noon and always blazing hot. There were none of the night animals—the bat, the owl, the kinkajou, the biting ants.

Without the presence of darkness it was difficult to sleep. Men and women worked unceasingly at their customary tasks: weaving hammocks from palm fiber, grating manioc roots, making arrows, spearing fish and hunting howler monkeys. In the earth's green youth, this pattern of life may have been satisfactory, but now the people were growing tired of constant labor.

In those days, gods and humans lived together, and existence was equally arduous for both. Then the Great Serpent's daughter—who, according to the legend, was married to a mortal man—decided to ask her father for some magical means of dissipating daylight, so that all humankind could rest.

The story does not explain why the woman herself could not visit her father;

there may have been some taboo that prevented her from traveling. So she asked her husband's three brothers to make the journey to the great anaconda's dwelling-place and petition her mighty parent on her behalf.

The brothers prepared for their very important mission. They painted their bodies with long stripes of black genipa dye, wove themselves cotton armbands decorated with the feathers of the blue-and-gold macaw and made headbands of palm fiber and harpy eagle plumes.

Then they took their canoe and paddled upriver to the deep, dark lake where the Great Serpent lived. They saw the sun glinting on his coils as he lay amid the roots of a great tree.

It was a strange and eerie place. No parrot squawked, no monkey chattered, there were none of the teeming jungle's familiar noises. One of the brothers whispered to the others that he was frightened by the silence. It could only mean that the anaconda had swallowed every living thing for miles around.

Trembling, the men approached the Great Serpent. His body, as thick as the tree trunk beside him, looped around the roots and in and out of the lake. There was no end to him. He was the color of sun-dappled mud and the darker spots along his coils looked like staring eyes. The harpy eagle plumes on the brothers' headdresses quivered as their bodies shook with terror.

The scaly lord gave them leave to speak and they conveyed his daughter's message. She begged some darkness, they explained, to give respite to the world's weary inhabitants.

The Lord of Creation heaved himself about until his head emerged, rising up from the thick coils of his body, to tower over the brothers. They clung to one another in fear: He was massive enough to swallow a full-grown man as easily as a man could swallow a tiny fish.

Then the Great Serpent rose up until his head touched the branches of a peach palm that formed part of the canopy of the forest, then higher still until the brothers could no longer see his face and his body itself seemed to be one of the tall jungle trees. Suddenly, with a soft sound of skin on skin, he folded his coils back down again.

In his mouth was a glowing orange nut, the fruit of the peach palm, ripe and ready for cooking. The snake lowered his triangular head and placed the nut gently before the brothers.

He told them that the nut was an object of great magic and terrible power. Enclosed within its shell was night, the source and fountainhead of all darkness and the antidote to endless day. He warned them that only a divine being would have the ability to control this force once it was released.

The brothers were instructed to carry the nut very carefully back to the serpent's daughter. No matter what happened, they were not to open it. If they did, disaster would follow, and they themselves would be severely punished.

With courteous words and gestures, they took their leave of the reptilian god.

They pushed their canoe out and paddled back downriver, happy to have escaped with their lives. As they traveled, they heard noises coming from the peach palm nut—cries, hums, buzzes, whines and scratching sounds.

The brothers were perplexed. Nuts, in their experience, did not make noises. They wondered if something was wrong with its mysterious contents. If they delivered damaged goods to their sister-in-law, she would hold them entirely responsible. They quailed to think of her terrible fury if their precious cargo turned out to be flawed in any way. Forgetting or perhaps ignoring the Great Serpent's warning, they decided to open the nut and investigate.

They paddled the canoe to the bank and tied up. They made a fire by twirling a dry twig against a fire-stick with their hands until a spark came. Then they heated water in a clay pot and boiled the nut to soften it. Finally they separated the halves of the nut.

Suddenly, with a sound like rain beating on the forest leaves, night flew out of the nut. With it came all the nocturnal insects: Mosquitoes swarmed around the brothers' heads, termites landed upon rotten tree trunks, ants covered the forest floor. Day was hidden by a cloud of humming insects and the entire jungle was obscured by blackness.

The insects swarmed through the forest, biting all the men who were out hunting. And by the powerful magic released from the nut, these innocent victims became the animals who forever after were to dwell in darkness. Some turned into night-monkeys, others became owls, the rest became bats which swooped through the clouds on shadowy and silent wings.

When the brothers reached home, the woman was waiting for them, her eyes blazing out of the all-encompassing darkness. News of the disaster had traveled ahead of them. Because of their intervention, the night had run wild.

The bearers of the darkness would have to be punished. Accordingly, the Great Serpent's daughter used her divine powers to transform her three brothers-in-law from men into night-monkeys.

They were sentenced to dwell in the forest, away from the fires of men, prey to all the ants and biting insects of the night. Because they had so stupidly disobeyed the great anaconda's command, they would henceforth be condemned to swing among the branches of the trees, chattering foolishly.

Using magic means known only to herself, the daughter of the Great Serpent imposed order on the untrammeled darkness that floated over all the world. She made the cubuju bird, painting white on its head and red on its feathers, and appointed it to sing every evening to summon the night. She created the inambu bird and powdered it gray with ashes; she assigned this bird to sing in the morning to call back the day. And from that moment onward there was a time of light for humankind to work, and a time of darkness for resting, loving and telling stories ☆

The Twin Luminants

According to the first interpreters of nature's mysteries, the sun and the moon were kin. They were sometimes thought to be husband and wife, and sometimes brother and sister. Many nations worshipped them as gods, members of a sacred family, who had always lived in the heavens, but others remembered them as ordinary human children who took refuge in the celestial orbs to escape some terrible fate on earth.

It was not surprising to find children in such dire straits that the sky provided the only place of safety. For the old lore was filled with chronicles of young people in mortal danger. Siblings, usually orphaned and friendless, were menaced by witches, stalked by wolves, besieged in their own homes by murderers, tormented by sadistic stepparents or simply abandoned to wander and starve in the forest. Their sufferings were the stuff of nightmares. But their eventual salvation—through their own quick wits or the benevolent intervention of supernatural forces—exorcised the demons of the dark.

A Korean legend told of two such imperiled orphans. They lived in an age when the skies were empty, and daylight came faintly and sporadically from an unknown source. It was a bleak and hungry era, with no summer heat to bless the crops and no certain means of marking the passage of seasons.

In those days the world was thickly covered by forests, where darkness and its creatures ruled. People clustered in small settlements and waged a constant war against the encroaching vines and undergrowth. They ate little and slept less, forever on their guard against the beasts that stalked them in the night.

The fiercest of all these predators were the tigers. They were known to rend human flesh with their teeth, to score bloody tracks with claws as sharp as razors and to maul people so badly that even their own families could not recognize them. But, worst of all, they possessed the power of speech.

Their ability to use language did nothing to humanize them. Instead it added a great and sinister weapon to the arsenal they already had. They were able to converse with their victims, using this power to raise the hope of salvation before dashing it forever with the cruel stroke of a claw. The gift of tongues in a cunning and carnivorous beast was open to vast and terrible abuse. And tigers killed not only for food, but for sport.

In the village where the Korean story took place, the peasants lived in a state of perpetual, mind-numbing terror. Tigers preyed on them even as they huddled together in their homes. No thatched roof or wall of split bamboo was strong enough to withstand the giant cats. Yellow eyes glowed amid the trees that pressed close to the settlement, and the tigers' roars filled the air.

The brother and sister who were the hero and heroine of the tale lived with their mother in a hut at the edge of the forest. Their father had been killed by a tiger and, ever since, their mother was obsessed by her fear of the creatures. Her talk was all of tigers.

She taught the children everything her people had learned about their most dreaded of enemies—their moods, their hiding-places, their seasons of breeding. She explained that every tiger was given a name, depending upon its character, size and age: Rampager, Old Scarface, Coward. The fiercest and strongest of the tigers, those in their prime and at the height of their awesome powers, were all known by the villagers as Kal Pem, or Fierce Heart. And it was one of the Kal Pem which had slaughtered their father. The mother walked in terror of meeting the same fate. This fear was contagious; her children cowered inside the hut, starting at every unexpected noise and suspicious of strangers.

Each day the mother's dread of what terrors lay in wait outside the door warred with her children's hunger and each day hunger won, forcing her out onto the dangerous forest paths to labor in the fields of a wealthy neighbor. As she groped her way through the forest, the woman imagined that she could smell the tiger's rancid breath in the thick gloom. Every bush, every tangle of vines, every clump of ferns seemed to her to hide a tiger's muzzle, wet and matted with blood and spittle.

One day, she crept stealthily back from the market in a neighboring village, carrying a tray of buckwheat puddings for the family's evening meal. Suddenly, a vivid streak of orange shot out from the bushes and the

In the darkest of all ages, when no sun or moon hung in the heavens, two
        children fled the ravening jaws of the tiger, dreaded creature of the night.
Taking refuge in the sky, they brought light to the world.

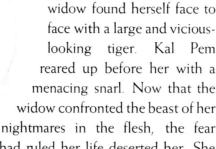

widow found herself face to face with a large and vicious-looking tiger. Kal Pem reared up before her with a menacing snarl. Now that the widow confronted the beast of her nightmares in the flesh, the fear that had ruled her life deserted her. She stood her ground, drew a deep breath and asked Kal Pem what it wanted.

The tiger thrust its slavering jaws into her face and replied that it was hungry. Quickly, she thrust the largest of the puddings at the tiger and ran off into the forest. As she fled, she tripped over roots and trailing vines, her heart pounding loud enough to be heard by the people in the next village.

It was not long before the tiger caught up with her, demanding more food. The widow, who was by now exhausted, gave it the rest of the puddings and rushed away as fast as she could, but the beast was upon her again in an instant. This time, Kal Pem asked for one of her arms to eat, threatening to kill her if she did not agree.

The storytellers say that the woman yielded up her arm without cry or complaint, and that the evil beast ate it while she made one last vain attempt to outrun it, leaving a trail of blood behind her. More cruel than hungry, the tiger followed her and demanded, one by one, the widow's other arm and both her legs. Finally, when Kal Pem had grown tired of the chase, it set upon its victim and devoured what remained of her body. Then it put the woman's bloodstained kerchief on its head as a grisly souvenir of the kill.

Its hunger for blood sharpened rather than sated by the meal it had just eaten, the tiger loped through the forest toward the village. The first house it came to was the widow's poor hut. The tiger walked around it once, sniffing carefully under the low eaves, pressing its nose against the bottom of the door and trying to peer through the oiled-paper window. Its keen ears made sense of the furtive rustlings and told it that there were two children within. The tiger could smell the fear of its victims trapped inside their home. It licked its lips. No delicacy on earth could be more pleasing than the tender flesh of human young.

The beast was hungry enough to want to eat again, but full enough to enjoy the sport of toying with its defenseless prey before the slaughter. Kal Pem called out to the children in a rasping imitation of their mother's voice, asking them to open the door and let her in.

The girl was not convinced by the accents on the other side of the door, and said so. The tiger's voice grew harsher as it demanded entry.

The boy wondered aloud why his mother sounded so hoarse. The counterfeit mother explained that she had been shouting all day long, to drive the crows away from the crops.

The boy was satisfied with this answer and wanted to open the door, but his sister refused until she had proof that the voice was her mother's. She opened the window, just a crack, and asked to see her mother's hand. Kal Pem was beginning to grow tired of the game, but it sheathed its claws and offered its paw.

When the girl reached out her hand and touched the rough, hairy paw, she recoiled. Then she heard her mother's voice, protesting that many hours of washing and starching clothes had made her arms rough.

The children peered under the door and, in the darkness, they could just make out the stripes on the tiger's coat. The game was over. The pair fled out the back window of the hut. Silently they climbed a high tree that overhung the house, hoping they would be safe from the intruder.

But the beast soon realized that the children were no longer in the hut. Circling the ramshackle structure, it noticed the open window at the back. Behind the house was the well from which the family drew their water. The tiger peered down into it, and saw the children's terrified faces floating on the mirrored surface. With a roar of hunger and triumph, Kal Pem lunged at the waters and scooped up the images with a lightning stroke of its paw. The faces wavered and disappeared, making the tiger angrier and more frustrated than ever.

Then it heard a rustling noise, looked up and saw the two children perched in the branches overhead. In one bound it was at the bottom of the tree and with another it had started to climb, digging its claws into the bark.

The terror the children had endured inside the hut was nothing compared to what they now felt as the tiger's weight shook the massive tree. They saw their mother's tattered and bloodstained kerchief on the tiger's head and realized at once what had happened. They began to pray to the gods to take pity on them in their orphaned state, and to save them from the tiger's jaws.

The storytellers said that the children's prayer was answered in an instant. The gods lowered a rope from heaven, and the two children quickly scrambled up toward the sky.

The tiger was right behind them. Sure-footed as any feline, it leaped from the treetop to the rope and began to climb. But its body was heavy with the undigested flesh of its most recent victim. The fibers of the rope strained, frayed and finally broke under its tremendous weight, sending the tiger crashing to destruction. Thus, even in her death, the widow had her revenge.

And now the gods in their wisdom decided that the time had come to end the rule of darkness and the tyranny of tigers. They endowed the children with supernatural brilliance and commanded them to shine in the heavens for all eternity. The girl became the sun, burning so intensely that no one could look directly at her. The boy reigned as the moon, throwing his light across dark, empty space to brighten the night. From that day forward, nightmares ended with the dawn. The creatures of the dark lost half their kingdom, and tigers never regained the power of speech ☆

The Light-Eaters

Whenever the sun hid behind a veil, or the moon suddenly vanished from a cloudless sky, the peoples of the old world trembled. They knew that eclipses were caused by the hungry monsters who perpetually stalked the celestial bodies: a howling wolf from the Scandinavian spruce forests, a sinuous Chinese dragon, a giant fish from the magical bestiary of the Jews.

Sometimes, when one of these creatures sprang at its prey and took a bite, the sky went dark. Then everyone knew what had to be done. They screamed, shouted, prayed, banged drums or set off fireworks, anything that would frighten the predator away. Thanks to their valiant efforts, the sun or moon would soon reappear, unscathed, in the sky.

The Fractured Moon

From the very beginning, the moon's mutations puzzled ordinary people and exercised the imaginations of the sages. Its wandering path through the night sky, its time of rising and setting, and above all its changing shape as it waxed and waned—swelling to gravid splendor and then dwindling once again to a humble shred—all demanded an explanation that could make some sense of the mysterious laws of the heavens. Was the moon object or spirit, male or female, human or animal, benign or malign? What caused its cyclical alterations?

One answer came from medieval Germany, a world of spreading dark forests interspersed with small towns run by hard-headed, hard-pressed folk, bound in a tight web of local loyalties and rivalries. Apparently, to these down-to-earth burghers, the moon was an object of civic usefulness, its social and commercial value expressed by the quantity of nocturnal light it provided. Indeed, the tale they told to explain its monthly changes depicted it as something akin to a celestial form of municipal street-lighting. And yet, in spite of this determinedly materialistic point of view, people could not quite shut out of their minds—or their legends—certain whispers of the wilder, more anarchic times of their distant tribal ancestors.

To the early Teutons, as to all the pagan races of old Europe, the moon was the wellspring of ancient affinities and forces which controlled fertility in both plants and humans, and perhaps drove unlucky souls to madness. And even after the tribal trading posts had turned into walled cities and the barbarian chieftains had become burgomasters and tax collectors, the old beliefs persisted and lurked under the surface of what appeared to be a simple story. In the chronicle in question, the moon was not merely a lamp for lovers; it was imbued, for a time, with the ominous power to wake the dead. This is how it happened.

There was once a land where the nights were always black. In this country there was no moon, and stars were unknown. When the sun sank each evening, no gleam of light was seen until the next day dawned.

There came a time when four friends from this dreary land decided to set out on their travels to explore the world. During their wanderings, the young men discovered a kingdom that was hitherto unknown to them. On the day of their arrival, as evening drew on and the daylight began to ebb, they took care to find a sheltered spot under a hedge where they would be able to spend the night in warmth and safety. They took it for granted that once the last light of day had died they would—just as in their own country—be unable to see their hands in front of their faces till morning.

But when the sun had set and the color of the sky had changed from pink to green and finally to black, they gradually became aware that, although it was indubitably night, they were nevertheless still able to see. To their astonishment, everything was palely visible, drained of color but clearly evident to their startled

At first the moon was merely a lamp hanging in an oak. Lured by its light,
the secretive Others—dwarfs, elves and their troglodytic ilk—emerged
from their hiding-places to dance until dawn.

 eyes. It was as if the thick darkness to which they were accustomed had been rendered transparent. They puzzled over the cause of this phenomenon. At last, as they cast about for an explanation, they perceived a large luminous globe suspended in a nearby oak tree. It shed a magical light further than any lamp, suffusing the lanes and farmyards with a soft glow as far as the men could see.

As the travelers stood marveling at the mysterious orb, they saw by its gentle light a horse and cart approaching. It was driven by a peasant of the district. He sang to himself as the wagon rumbled over the ruts, apparently unruffled by the fact that he was driving home after dark—something no one would dream of doing in their own land.

The young men stopped the rustic and asked what sort of light it was that hung there in the tree. With the natural courtesy of countrymen, he concealed his astonishment at their ignorance. He told them that the object in question was the moon, which the mayor of that place had purchased from some unknown supplier for the sum of three talers. It was the mayor who had hung the moon up in the oak tree, and the mayor who was responsible for operating it. Every day he refilled it with oil and kept it clean and trimmed so that it burned steadily with a clear light. For this service, the community paid him a fee of one taler a week.

When the cart and its driver had gone on their way and were safely out of ear-shot, the travelers conferred with the greatest excitement about their discovery. They could easily see what an advantage such a lamp would be for their own citizens at home. Their imaginations were fired with visions of what life might be like if night were accompanied by this benevolent glow instead of the terrors of pitch blackness. It must also be said that the economics of the proposition appealed to them. They calculated that the mayor of this place recovered his initial investment regularly every three weeks. But the travelers were not contemplating any financial outlay for the lamp at all. They were going to steal it.

In an instant, the most agile climber among them was up in the top branches of the tree, where he unfastened the shining sphere and lowered it carefully to his accomplices on the ground. When the moon lay at their feet, they muffled its incriminating light with their thick black cloaks. In all the houses nearby, doors and windows flew open at once. Neighbors shouted to one another in the unfamiliar darkness and stumbled out into the night to look for their moon. But the thieves, bred as men of moonless nights, were able to elude their blind pursuers and found their way to the borders of their homeland before dawn. They took their plunder to the tallest oak tree in the country, and hung the moon high up in its branches.

It was significant that the lamp-post had to be an oak. Ancient societies

 regarded the tree as hallowed and awe-inspiring, the symbol of godhead, if not a god itself. Some said the roots of the oak went down to the underworld, others that it was the tree of the dead and the home of departed spirits; still others believed that its leaves sighed out the secrets of the future. Oak groves were held to be the sites of sacred rituals, even—it was whispered—of human sacrifice. Certainly no other tree could claim a better right to carry the moon through the night.

As the moon shed its light over the country, the silvery beams penetrated all the houses, flooding bedrooms and kitchens with a gentle glow, to the delight of young and old alike.

The light woke other beings too: Dwarfs came blinking from the musty depths of their caves under the mountains, and elves who smelled of leaf-mold and mushrooms danced all night in the meadows, leaving blighted circles that made the peasants stare in wonder when they came to their work in the morning.

Once the moon was installed, the four friends negotiated a contract with their fellow townsfolk along the same lines as the scheme worked out by the moon's previous owner. The men agreed to see to its maintenance in just the same way, keeping it supplied with oil and trimming the wick, and every week they shared the fee of one taler for their trouble.

For many years the arrangement continued unchanged, to the satisfaction of all concerned. The moon's licencees

 grew wealthy and spent freely on life's pleasures. But as the four men declined into old age, they became crabbed and miserly. Eventually one of them fell ill and, realizing that death was near, he stipulated that his share of the moon—one quarter—should be buried with him.

When the man died, the mayor of the town, honoring his final instructions, climbed the oak tree himself and, with a pair of hedge clippers, sheared off one quarter of the moon, which he faithfully nailed into the coffin with the corpse. The moon's light, although partially diminished by the loss, continued to illumine the nights.

Soon afterward, the second man died, taking his share of the moon with him in the same way as the first and leaving the orb with only half its original strength. When the death of the third man occurred, the light became even weaker, for he too had required that his share should be cut off and buried with him in his coffin. Finally, with the funeral of the fourth and last of the men who had brought home the moon, the light that the people had become so used to vanished completely.

But down in the nether world of the dead where the four men had gone, the lunar fragments, drawn by their own magical magnetism, had joined themselves together again. Now the full moon shone in the place where total darkness had always reigned before. As the strange light glimmered around them, the dead stirred restlessly and began to wake. To eyes weakened by the unbroken darkness of the world below, the gentle moonlight seemed as bright as day.

Thus unexpectedly aroused, all the dead sooned regained their strength and their good spirits. They rose up and immediately made their way back to the land of the living. Some followed dimly remembered roads to find homesteads much altered by the years and grandchildren grown to old age. Others set out to re-open old quarrels which had once been settled by death. The most painful discovery for many of them was how completely they had been forgotten, and how those still living feared and shunned them. The bitter and quarrelsome members of the undead soon met in the taverns, where they found their more cheerful companions toasting their re-awakening with glass after glass of the wine they had loved in life.

It was not long before the formerly dead ran riot. The drinkers did not know when to stop, and the sweet wine, so long untasted, fueled their tempers. A nudge became a shove and soon they fell to blows. Overturning tables and splintering chairs, they began to fight in earnest. The noise and hubbub grew steadily wilder, until it seemed as if the apocalypse had come.

The gods and angels who ruled the universe grew angry. The rising-up of corpses violated natural law, and the disturbances in the lower world seemed to signal a serious rebellion of its inmates.

The possibility of an uprising from the graves where many centuries of dead lay had long been foreseen in heaven and precautions had been taken against it. Armies of celestial troops had been formed, and waited in readiness to drive back the forces of evil, should they ever dare to storm the abode of the immortals. The invasion did not materialize, but the confused chaos in the world of mortals below continued unabated.

Finally a delegation of heavenly messengers, mounted on winged horses, rode out through the gates of heaven. They made straight for the frenzied fleshpots where the dead were creating havoc. Their authority proved quite sufficient to restore the world to its accustomed order and to bring the unruly dead back into subjection. Cowed by the presence of the immortals, they were persuaded to lay down again in their graves. The celestial troop returned to the heavens and took with it the moon that had so effectively roused the departed from their long sleep. They suspended it in the sky, a place of safety.

There it hung forever after, illumining not one small country or single municipality but the whole world. Yet it still, perhaps subversively, broadcast its message of rebirth and resurrection. At regular intervals, one quarter faded away or drifted off, then another. But always, the quarters reconstituted into a round, full moon, conveying the message that life was not a single, short, straight line but a never-ending cycle ☆

When the four men who owned the moon reached the ends of their mortal spans, each took his quarter share with him to the underworld. There the gleaming orb, reassembled, woke the dead and set them celebrating.

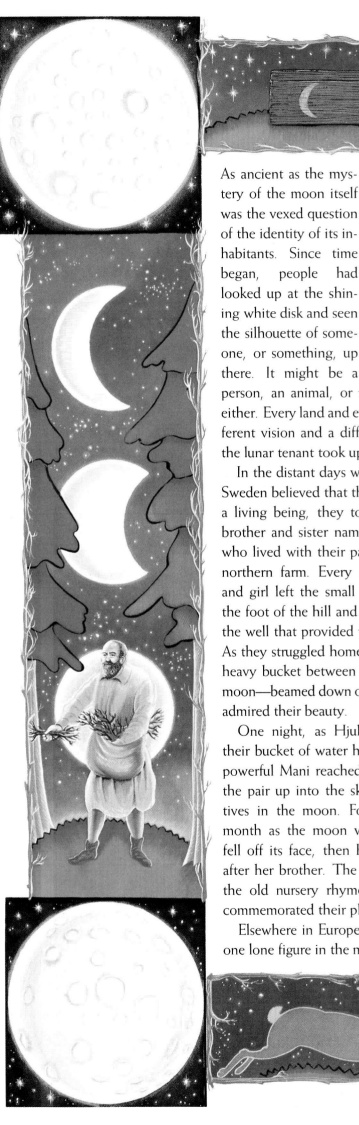

Lunar Denizens

As ancient as the mystery of the moon itself was the vexed question of the identity of its inhabitants. Since time began, people had looked up at the shining white disk and seen the silhouette of someone, or something, up there. It might be a person, an animal, or more than one of either. Every land and every age had a different vision and a different tale of how the lunar tenant took up residence.

In the distant days when the people of Sweden believed that the moon itself was a living being, they told the story of a brother and sister named Hjuki and Bil, who lived with their parents on a lonely northern farm. Every evening, the boy and girl left the small wooden house at the foot of the hill and climbed a path to the well that provided the family's water. As they struggled home carrying the full, heavy bucket between them, Mani—the moon—beamed down on the children and admired their beauty.

One night, as Hjuki and Bil carried their bucket of water homeward, the all-powerful Mani reached down and swept the pair up into the sky, to live as captives in the moon. Forever after, each month as the moon waned, first Hjuki fell off its face, then Bil came tumbling after her brother. The tale traveled, and the old nursery rhyme of Jack and Jill commemorated their plight.

Elsewhere in Europe, people saw only one lone figure in the moon—an old man carrying a load of firewood. According to some accounts, he was an earthly thief who had stolen the sticks from a neighbor's woodpile; others say he was a poor peasant who came by them honestly. But in either version of the tale the bare facts remained the same. An old woodsman, who lived in a village in Germany, was seen carrying a bundle of sticks on a Sunday. Such labor was forbidden on the week's sacred seventh day, which the ancients dedicated to the solar gods, and their descendants reserved for Christian devotions.

On this particular afternoon, when his neighbors were at their prayers, the old man trudged out of the woods carrying his burden of firewood. At a crossing of two paths, he met a stranger dressed in a fine suit of holiday clothes.

"Do you work on Sundays, old father?" the stranger asked, scandalized at the sacrilege. The woodsman laughed. "Sunday, Moon-day, it's all the same to me. I don't care. I never rest."

"Then never shall you rest," replied the stranger as, with a wave of his hand, he changed into an angel. "Every day for you shall be a Moon-day." And, with a slight nod of his head, both he and the woodsman vanished.

When the villagers finished their worship that afternoon, they soon noticed that the old man was missing. The peasants organized a search, and they joined

together to comb the surrounding woods. But no trace of the old man was found. As darkness fell, the men returned to their homes. And then the moon rose in the evening sky. Looking up, the villagers gasped with astonishment. There, in the

moon, they saw their neighbor, his back bent beneath the bundle of sticks slung across his shoulder, toiling for eternity.

While in Western eyes the moon had been home to human forms, in the East it had long been seen as the eternal resting place of a wild animal. Many thousands of years ago, the Buddha roamed the earth, disguised as a humble peasant. On one of his travels, he came upon an ape, a fox and a hare, that lived together in a clearing in the jungle.

"I'm hungry," said the Buddha to the three animals. "Won't you find food for a poor old man?"

So touched were they by the peasant's obvious need, the animals immediately set themselves to the task of gathering a feast. The ape scaled the trees, picking the sweetest fruit and meatiest coconuts he could find. The fox, with his sharp wits and pointed nose, poked among the undergrowth and sniffed out plump mushrooms and fragrant herbs.

They carried their harvest back to the clearing and spread the bounteous meal before the old man. He, meanwhile, had gathered a pile of twigs and branches together, and now he set them ablaze to make a fire with which to warm himself and his animal friends as darkness began to descend. At that moment the hare straggled into the clearing, panting with exhaustion. Not having the agility of the ape or the cunning of the fox, he had been unable to find any food to offer the peasant. Distraught at his inhospitality, he bowed before the old man.

"I have found nothing to give you, sir, to ease your hunger. Please let me offer you my humble life." And with that the hare leaped into the flames.

In an instant, the old man shrugged off his ragged clothes and stood tall and splendid as the Buddha. He reached into the fire and snatched the dying hare from the glowing embers.

"To lay down your life for a man in need is the greatest sacrifice any creature can make," spoke the Buddha. "In gratitude, I shall free you from death eternally." And the Buddha lifted the hare up into the moon for all on earth to see.

Some storytellers said that the creature lived up there in a shining box, secured by a lid, which the hare opened a little more each day as the moon progressed through its quarters. Others said he lived in a crystal mansion with fifteen windows. As the new moon began to wax, the hare, resplendent in a coat of silver, opened his windows one by one each day, until the light shone forth from all of them, revealing the full moon's glory ☆

Lindu's astral veil

The bards of Estonia told a sad tale of a goddess named Lindu, daughter of Uko, who was King of the Heavens. Lindu was a sovereign in her own right—Queen of Birds. She lived on the shores of the Baltic Sea among her winged subjects. In every season, she knew where each bird belonged and made certain that they kept to the right paths as they migrated across the skies.

Many suitors came down from the heavens to woo her. Lindu rejected the Sun and then the Moon, finding them altogether too predictable; every day, they followed the same course through the heavens. She dismissed the Pole Star as too set in his ways, never leaving his post in the sky. But when she was courted by the Northern Lights, she was so enthralled with his beauty and splendor that she fell in love at once. The Northern Lights could not stand the brightness of day for long, so, after they were betrothed, he returned to the land of midnight, promising to return for the wedding.

Lindu happily prepared for her marriage, but the Northern Lights proved himself as inconstant as he was beautiful. He never came back. Day after day, Lindu grew sadder and sadder, until her wedding dress was soaked with tears. She wept in the meadows, heedless of the fluttering birds that gathered round her.

Uko looked down from his palace in the heavens and took pity on his daughter. He commanded the winds to take her away from the earth and place her near him in the sky.

Once there, her wedding veil blew out behind her, and its threads were turned into a million glistening stars, their lights forming the night-pathway called the Milky Way. She continued to watch over the migrating birds, said the storytellers. Sometimes, on a winter's evening, she glimpsed the elusive Northern Lights, but that old grief was forgotten and she wondered why she had ever loved him.

Hieroglyphics of the Heavens

Stars in their millions, piercing the sky, were mysterious and, like the night itself, more than a little frightening to people of old. But these beads of celestial fire were not randomly scattered across the heavens like rice grains spilled from a broken basket. The stars formed patterns, divinely preordained, that carried messages and meanings.

Gazing upward, people could see the bright outlines of beasts, heroes, maidens, weapons, jeweled diadems and other wonders. They knew that each of these objects was there for a reason, even if the explanation for its presence differed from one land to the next.

The Greeks, living under the limpid Aegean skies, became as familiar with the stellar landscape as they were with their own herb-scented hills and islands. In the heavens they read the histories of gods and men—and a tangled, often violent history it was.

Victims of divine displeasure formed part of the starry population. Among these figures of the firmament was the dragon Draco, whose fifteen stars traced a tortuous coil through the northern sky. Its misfortune was to ally itself with the losing side in a war between the cosmic powers. Here is how it happened:

After the defeat of the Titans, their brothers the giants—the monstrous sons formed from the blood of Uranus—took up the battle against the new generation of young and vigorous gods and goddesses with the forms and faces of humankind. The giants had a hellish appearance— huge bodies resting on thick, sinuous legs ending in snakes' heads instead of feet.

For many years, the war turned all Creation into a battlefield. The sky was filled with a terrible roaring, as wounded monsters groaned and thrashed in their death throes. The giants, with a strength to equal their size, were nearly a match for Zeus and his companions: his consort Hera, his brothers Hades and Poseidon, his sister Demeter and his daughter the all-wise Athena. It was Athena who most terrified her adversaries: Her farsighted intelligence, her mastery of strategy and tactics, proved far more frightening than mere brute force.

Urged on by anger, however, in the heat of the battle, Athena displayed her divine strength by snatching an immense dragon from the ranks of giants. The great beast flew toward her, coiling and flexing, hissing malevolently, flashing its fangs. She raised her magic shield, cunningly crafted by Hephaestus, blacksmith to the gods, to deflect the dragon's exhalations of poisonous steam. Enveloped in a cloud of its own destructive vapors, Draco, scalded and stupefied, landed in a heap at Athena's feet.

Barely lifting her eyes from the battle raging before her, Athena scooped up the creature in one hand, swung it around her head and flung it up into the sky. Its long body crashed into the roof of heaven and stuck there for all time, marked out by the sinuous track of its stars.

On either side of Draco's thrashing tail stood the two Bears, the Great and the Little. They too were earthly creatures that found themselves in the sky as a

The fugitive stars

In New Guinea, legend tells of a time when the stars lived on earth. They were shy creatures and kept themselves well hidden from the eyes of men. In the deepest jungles, far from human habitation, they canoed along rivers overhung by creepers and branches. They wore glistening decorations of curving white boars' teeth and gleaming shells, and the glow given off by these ornaments wavered rhythmically in unison with the haunting songs the stars sang as they paddled.

One night a traveler, lost in the forest, saw the shimmering light and heard the mysterious music. Fired with curiosity, he approached the source of these marvels, and peered from behind a curtain of vines. He was nearly blinded by the magnificence he saw before him on the waters of a river. Perhaps he was greedy, perhaps he merely longed to hold such beauty in his hands; but on an impulse he reached out from his hiding place and touched one of the canoes.

Frightened, the stars escaped, took refuge in the skies and never returned. Their glorious brilliance was still visible from earth every night, but their songs never again charmed human ears.

result of divine intervention. After the old gods were deposed and the new regime safely installed on Mount Olympus, the king of the gods, Zeus, was able to turn his attention to more peaceful pursuits. He was a vigorous, lusty monarch, and well aware that his power made it possible for him to possess—by force or softer suasions—any female, divine or mortal, who attracted his roving eye.

One of his many loves was the nymph Callisto, a skilled hunter who spent her days pursuing wild boar, stags and other game in the forested mountains of Arcadia. With its impenetrable thickets, mossy glades, overhanging rocks and secluded valleys, this wild country provided a most congenial setting for chance meetings and romantic assignations. Zeus took full advantage of the opportunities. Day after day, he descended from the heights of Mount Olympus to enjoy the pleasures of the chase, with the nymph as his chosen quarry.

The goddess Hera, Zeus's consort, raged and seethed at what was merely the latest in a long line of her husband's infidelities. She was tired of finding twigs of Arcadian greenery ensnared in his garments, and she stopped up her ears when she caught him absentmindedly whistling a poor imitation of the notes that Callisto trilled on her hunting horn.

Powerless against her immortal, philandering spouse, Hera instead directed her jealous wrath against Callisto. One morning, once she was satisfied that Zeus was safely asleep after a night of Arcadian revelry, the goddess made her way to the forests from which her lord had lately returned. When she found Callisto in a clearing, trussing up a slaughtered deer, Hera introduced herself, made a few disparaging remarks about her husband's taste in women, and swiftly transformed her rival into a shaggy bear.

To add piquancy to this punishment, Hera left Callisto's human emotions and soul intact within her new ursine body. As a result, Callisto roamed the forests in terror, not only fearing the hunters— once her comrades, now her enemies— but dreading the savage beasts in whose wild company she was henceforth forced

Flung into the air by an angry goddess, the dragon stuck fast to the roof of heaven.
Fifteen stars mapped out its sinuous coil and outstretched tail.

to live. From her vantage point on Mount Olympus, Hera watched Callisto's sufferings with sour satisfaction.

One day, the goddess's perverse pleasure was heightened when she saw a young hunter approaching the bear. She recognized him as Callisto's son, Arcas, fathered by Zeus and now grown to handsome manhood. Hera observed that Callisto, too, appeared to recognize her child. The bear, forgetting her own outward appearance, rushed forward to embrace her long-lost offspring. But Arcas, seeing only an agitated beast running toward him, raised his spear for the kill.

To Hera's chagrin, Zeus chose that moment to look down from Olympus. Promptly, he stopped the young hunter as he began to hurl the spear and changed him into a bear, somewhat smaller than his mother. Then he grasped both bears by their tails and hoisted them into the

sky to a place of safety among the stars.

Hera, purple with fury, left Olympus and went to cool her ire with a visit to the damp realms of the sea-god. She told him how her unfaithful partner had offended her by first dallying with Callisto and then elevating her and her son to an exalted position in the heavens.

Sympathetic to Hera's marital woes, the marine god promised that he would never allow that upstart female, or her bastard, to slake their thirst in the oceans. And for that reason the constellations of the Great Bear and the Little Bear always remained high in the sky and never dipped down below the horizon into the cooling waters of the sea.

The sea-gods, it seemed, were instrumental in the placement of another starry configuration, that of the lady Cassiopeia who sat majestically upon her throne on the opposite side of the North Star from the Great and Little Bears. Her story began in ancient Ethiopia, at that time a happy and prosperous kingdom ruled by the monarch Cepheus. The lady Cassiopeia was his queen.

Cepheus reveled in his wife's beauty. The court poets sang of it, and the sculptors of the realm tried tirelessly to capture her exquisite features in stone or marble. But neither lyre nor chisel could do anything but hint at the breathtaking magnificence of the lady herself. Their failure to portray perfection may have disappointed the artists, but it surprised no one, least of all Cassiopeia, who was not burdened with any false modesty.

The Queen spared no effort, time nor expense when it came to her appearance.

Every morning, in the chilly hours before dawn, slave-women would assemble in her apartments bearing great armfuls of gowns and kirtles, robes and chitons. Each garment was woven in the softest fabrics, adorned with the most exquisite stitches, colored with the finest dyes.

Like a general reviewing the troops, Cassiopeia would pace up and down before the burdened slaves, bidding them unfold each costume for her closer inspection, casting those that displeased her into a bedraggled heap on the marble floor. And if the sun was high in the sky before she made her final choice of apparel and settled upon a suitable selection of jewelry, then so be it—for beauty, as she never tired of reminding her servants, deserved the taking of infinite pains.

Once dressed, Cassiopeia would join her husband, graciously accepting his compliments on her appearance. Then the Queen's young daughter, Andromeda, would be allowed to kiss her mother's hand. No closer embrace was

permitted lest the carefully chosen costume be marred or rumpled by some sticky excess of childish affection.

Attended by her maidens, the Queen took the air in the gardens of the royal palace, pausing in her promenade to contemplate her image in the reflecting pool. Then, after the ceremonious unlatching of a hidden gate, the entourage enjoyed a walk in the surrounding woodlands. From time to time, Cassiopeia would pause to examine a bead of moisture clinging to a leaf or a flower petal, to see if her reflection was visible. After an hour or so of this botanizing, when the late morning sun waxed hot, the party ambled down to the seashore to allow the Queen to bathe. Her favorite beach, reserved for her exclusive use, lay in a sheltered cove half-hidden by great rocks.

One morning, before plunging into the water, she paused for a moment to let

*Perseus slew the ghastly Medusa, whose gaze turned her victims to stone. In the hero's
stellar outline, he still carried the monster's head. But even in death she was dangerous,
and the brightest star of the whole configuration was said to be her lethal, staring eye.*

the sea and sky look upon her unclothed loveliness. Then, in a voice filled with certainty, she challenged the heavens to produce a goddess, or the oceans a sea nymph, half as ravishing as she.

The Queen was cheerfully indifferent to the presence of the nymphs of the water—the Nereids—in the neighborhood. If they were in earshot when she made her boast, it mattered not a whit to her. But it must have mattered to the Nereids, for, after she had spoken, she felt perfumed breezes rush past her. Instantly, a whirlpool appeared where the bevy of water nymphs had plunged down to their subaqueous home.

Unruffled, Cassiopeia shrugged her perfectly formed shoulders and proceeded with her swim. When at last she strode forth from the waters, her body sparkling, she was puzzled and displeased by the sight that met her. Her waiting handmaidens stood frozen on the shore, their jaws gaping slackly, their eyes round with horror and, worst of all, the towels and garments they held ready for her trailing carelessly in the sand. She began to scold them in a shrill voice, but then realized that the maids, unaware of her presence, were staring fixedly at something in the sea behind her.

She turned, chilled by a sudden shadow across the sun, to discover a monster— shaped like a whale but the size of a floating mountain—glaring down at her with an unblinking and malevolent eye. With a speed that belied its bulk, it moved out of the water and up the beach.

Snatching her robe from the limp grasp of a petrified servant, Cassiopeia abandoned all thoughts of custom, pomp and protocol and ran from the beach, her handmaidens hard at her heels.

Within hours the whole coast was devastated, the population fled to the hills in panic, and the palace, although safe from harm on its wooded promontory, was in an uproar. Inexorably, deliberately, the monster oozed and rolled its way across the landscape. Cassiopeia, still quaking from shock, and her husband Cepheus listened in horror to each fresh report of olive groves ruined, vineyards ravaged, farms laid waste and villages ground to dust.

The King's soothsayers and councillors were helpless; they confessed that it would need more than mortal wisdom to defeat the monster. The royal couple had no alternative but to don the simple, homespun vestments of mourning and humility and climb, painfully and barefoot, to the remote mountain cave of the Oracle, who served as a conduit for communication with the gods.

In the cell of the seer, deep within the mountain's heart, the smoke of incense and animal sacrifices stung Cassiopeia's eyes and sent King Cepheus into spasms of coughing. Upon hearing the questions and appeals they put to her, the Oracle fell into a fit, then lay still as a corpse. For a long time, the chamber's silence was broken only by the rustling sound of the sacred python slowly drawing its coils along the stone floor.

The Oracle rose up, her eyes glowing, and spoke in a voice that was not her

Cassiopeia, Queen of Ethiopia, sat in state in the sky, still on her regal throne. But the pride that was her gravest sin in life was punished after death, for her constellation spent half of every night turned upside down, reminding the world that even royalty was not exempt from comeuppances.

own. After her brief utterance, she fell once more into a deathlike trance.

Bowed down with horror, Cepheus and Cassiopeia stumbled from the cave. At its mouth they stood for a moment, blinded by the sunlight. Cassiopeia began to shriek and wail. Cepheus, racked with sobs, led his queen back to the palace. There was grim work to be done.

The Oracle's command had been unequivocal. An offense had been committed against Poseidon, god of the ocean, who had sent the horrible whale-monster Cetus as a mark of his anger. Appeasement could take only one form: a human sacrifice. But the crime had been committed by Cassiopeia, whose boasting had so belittled the sea-god's nymphs. And, as a result, only one victim would assuage the deity's wrath: It had to be Andromeda, Cassiopeia's virgin child.

The Oracle's instructions were precise. The girl was to be chained to a rock overlooking the beach where her mother had bragged of her own beauty, and left there until the monster came to devour her. This done, divine honor would be satisfied and the Ethiopian coast would be left once more in peace.

The next morning, as Cassiopeia looked on from a distance, Andromeda was led to the beach by palace guards and bound fast to a rock with iron chains. The guards then retreated to high ground and awaited the coming of the whale-monster to claim its prize.

For a while, nothing happened. Sea birds wheeled overhead, the waves lapped the shore, the salty breezes played, as if this were an ordinary day by the seaside. Only the motionless figure of Andromeda, wrapped in chains, struck a jarring note in the peaceful scene.

Then Cassiopeia, through her tears, saw that her daughter was no longer alone. A young man, golden as the sun, mounted on a great white winged horse, approached her across the sands. With a graceful leap, he dismounted and stood beside her. Cassiopeia was too far away to hear their discourse, but she saw him stroke Andromeda's forehead with a reassuring gesture. Then he spun round on his heel for, with a great boiling and churning of waters, the monster Cetus rose up out of the sea.

The stranger rushed toward the beast and plunged his sword into the creature's slimy body. But the monster ignored the thrust as if it had been the merest fleabite, and advanced toward Andromeda. Her defender raised his sword again and again, slashing the monster so deeply that its blood gushed in torrents and stained the shore crimson.

The onlookers screamed with one terrified voice as the stranger slipped and lost his footing in a pool of gore. Cetus advanced upon him, jaws gaping. The man leaped to his feet again. Snatching an object that hung from his belt, he thrust it into the face of the monster. The monster stiffened and, in the interval between two heartbeats, turned to solid stone.

Cassiopeia and Cepheus rushed down to the rock where the unknown hero was busily freeing their daughter from her chains. Forgetting, in their gratitude, the

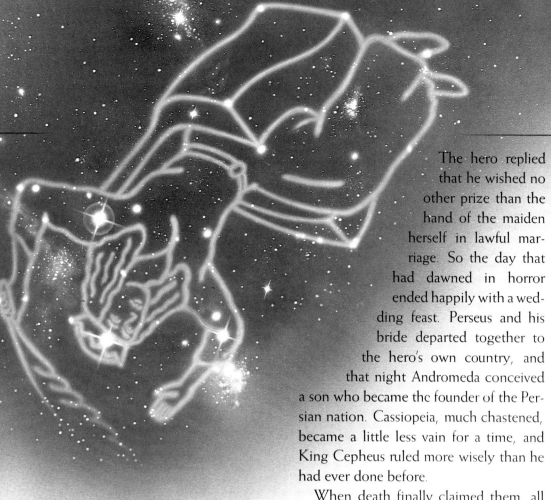

The hero replied that he wished no other prize than the hand of the maiden herself in lawful marriage. So the day that had dawned in horror ended happily with a wedding feast. Perseus and his bride departed together to the hero's own country, and that night Andromeda conceived a son who became the founder of the Persian nation. Cassiopeia, much chastened, became a little less vain for a time, and King Cepheus ruled more wisely than he had ever done before.

When death finally claimed them, all the players in the drama were honored by the gods, who gave them immortality as constellations in the sky: Cassiopeia, Cepheus, Perseus, his winged horse Pegasus, Andromeda and even the monster Cetus who, vainly, still pursued her lustfully through the heavens.

But the Nereids, the insulted sea nymphs, were outraged to see Cassiopeia rewarded in this way, even more so since Perseus had killed the whale-monster and cheated them of the revenge they had planned. Accordingly, the nymphs' protector Poseidon arranged for Cassiopeia to have her comeuppance; and so it was that, for all eternity, half of every night she could be seen sitting proudly on her celestial throne as it swung around the polestar. But for the other half of its orbit her chair was turned on its end, and the haughty Queen was forced to dangle, most humiliatingly, upside down ☆

barriers of royal etiquette, they knelt before him and wept with joy.

He raised them up and saluted them with the deference due their status. He introduced himself as Perseus, son of the god Zeus and a mortal princess. By a stroke of good fortune, or the kind intervention of some benevolent god, he had been riding his winged steed above the Ethiopian coast and had seen Andromeda chained to the rock. He explained that he had been homeward bound after carrying out a mission to slay the snaky-haired monster Medusa, whose ghastly countenance turned living beings to stone. Indeed, the creature's severed head, still writhing with serpents, had been the instrument of Cetus' destruction. Now it reposed again at Perseus' belt, safely concealed from view in a bundle of cloth.

The King asked Perseus what reward he craved for his rescue of their daughter.

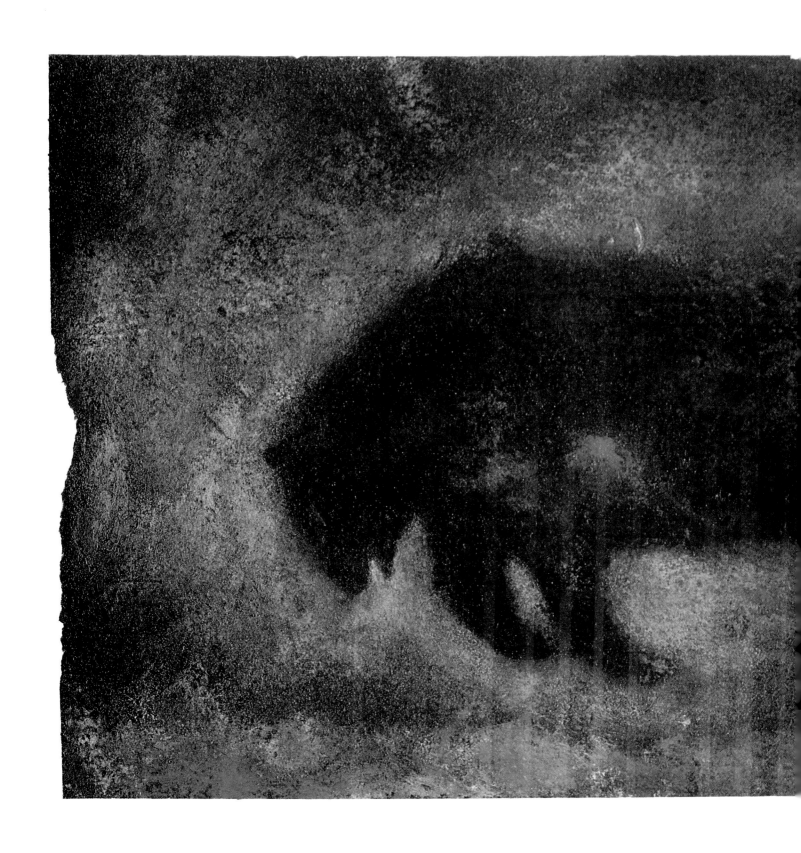

Dark Dramas of the Firmament

The ancients knew full well that the sky's changes and the caprices of the weather were the handiwork of gods, demons or other supernatural beings. Look upward, they said, and you will see games and rites and dramas. So it was that year after year, when American Indians were resting from the hunt by their campfires, they observed the astral chase of the Great Bear in the northern

skies. Throughout the spring, three hunters in the guise of single stars
stalked the celestial beast, who wandered the heavens unawares. By
summer, the constellation spun about the polar skies, for the quarry had
scented its pursuers and the chase was on in earnest. The hunters' arrows
found their mark in autumn. As the Great Bear's blood flowed it washed
the trees with crimson. The bloodstained leaves dropped from their
branches in mourning and the year died with the Bear. Its soul rested
through the winter, only to be reborn the next spring.

The long winter months were a time of darkness for the Eskimo tribes of the northern wastes. The sun glimmered for only a few hours each day, game was scarce, the sea was frozen and death was always near. Yet, in the dusky gloom that hung over the ice fields even in the middle of the day, the skies would sometimes glow with the shifting colors of the northern lights. The Naskapi Indians saw in those strange and beautiful lights that shone from the top of the world, leaping

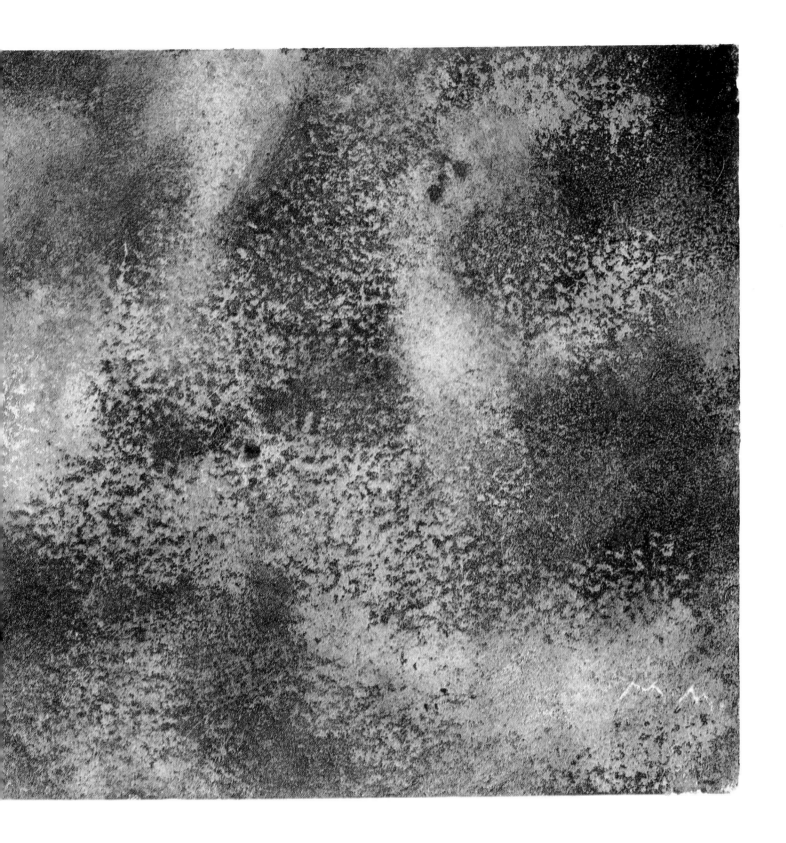

prism-like across the snow-covered tundra, a sign of life beyond death.

The world of the dead was a land of opposites, a mirror image of the earth where spirits walked upside down and cavorted on their heads. The aurora borealis was the window through which the living could watch the dead at play. Just as the tribesmen danced around their campfires after a hunt, so the souls of the departed danced in the heavens, casting their ethereal light onto the dark face of the world.

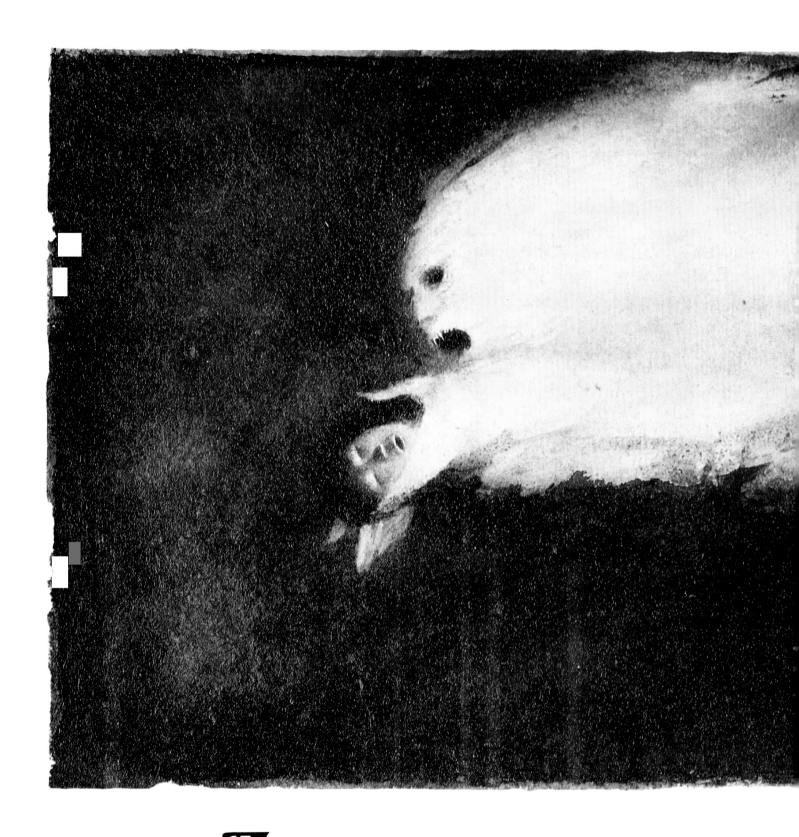

Another view of the afterlife prevailed on the wind-swept steppes of Siberia, where the spirits of the dead sought an unnatural return to life. There it was thought that dead souls were reborn as celestial vampires, thirsty for the human blood which alone could warm and sustain them. The vampires lurked high up in the skies, among the stars, waiting for cold, clear nights when they fell as meteors upon the silent earth, to

drink from the veins of the slumbering nomads. But vigilant folk, watching
the inky heavens, had the power to destroy the vampires before they
landed. The bloodsuckers were terrified by the sight of wakeful humans.
Shaken off their deadly course, the shooting stars fell harmlessly to
earth somewhere beyond the horizon, where their awful shrieks
resounded in the emptiness.

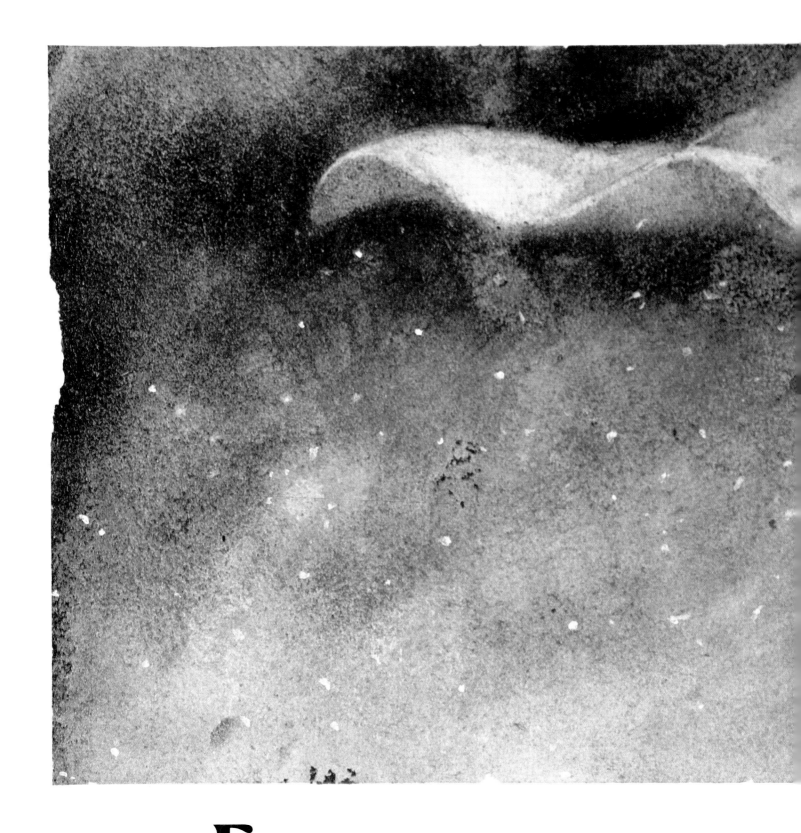

Kunchbacked, hook-toothed Mother Holle, ancient goddess of the
sky and the wind, queen of witches and elves, stumped about the
mountains and meadows of Germany with her cat and her crutch, perhaps
in search of her sacred herb, hellebore. Mortals lived in constant fear of
meeting her, for she could strike the sight from any man's eyes or the very
reason from his brain; she could shrink or swell at will—appearing

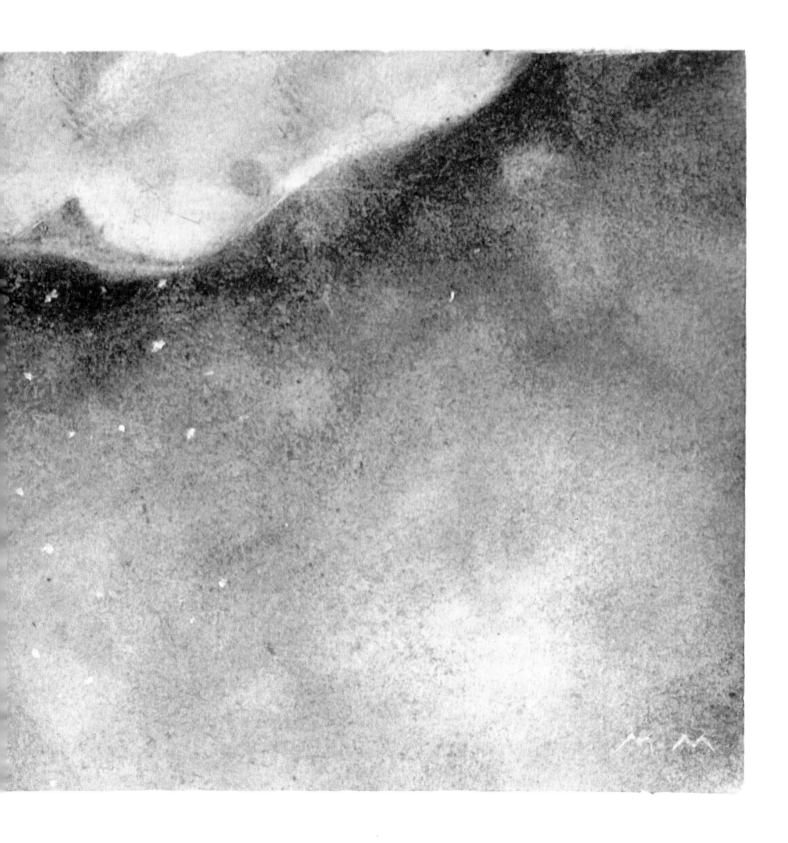

sometimes as a goblin, and at other times as a giantess.

Her pleasures, though, were homely. She liked to light a warming fire and sit spinning by the hearth. She sent gifts of gold to mortal women who tended their houses well. And, when winter snows flurried outside their windows, the women peered up into the clouds to look for Mother Holle, shaking out her feather bed.

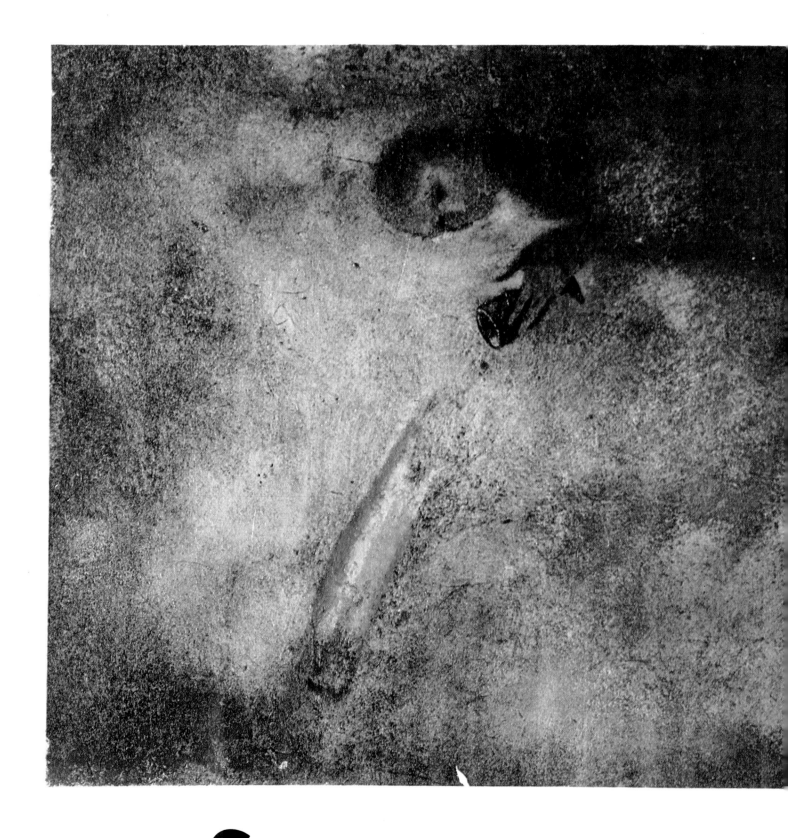

The Shans, dwellers in the Burmese mountains, worshipped a pantheon of divinities—gods of the sun, the moon, the clouds, the air. The sun, the moon and the stars were the golden palaces of these deities. When they visited earth, they made their home on Mount Meru, which itself rested on the back of a great fish. Human in their appearance, passions and follies, these deities ruled all the spectacles of wind and

weather seen on earth. The Shans could always tell when the gods had been reveling at a banquet in their home atop the sacred Mount Meru. As they feasted on succulent meats and luscious fruits, servants continually filled their glasses with celestial wine. In their gaiety, the gods overturned their goblets, spilling wine down through the clouds and leaving behind a vivid trail of color—a rainbow.

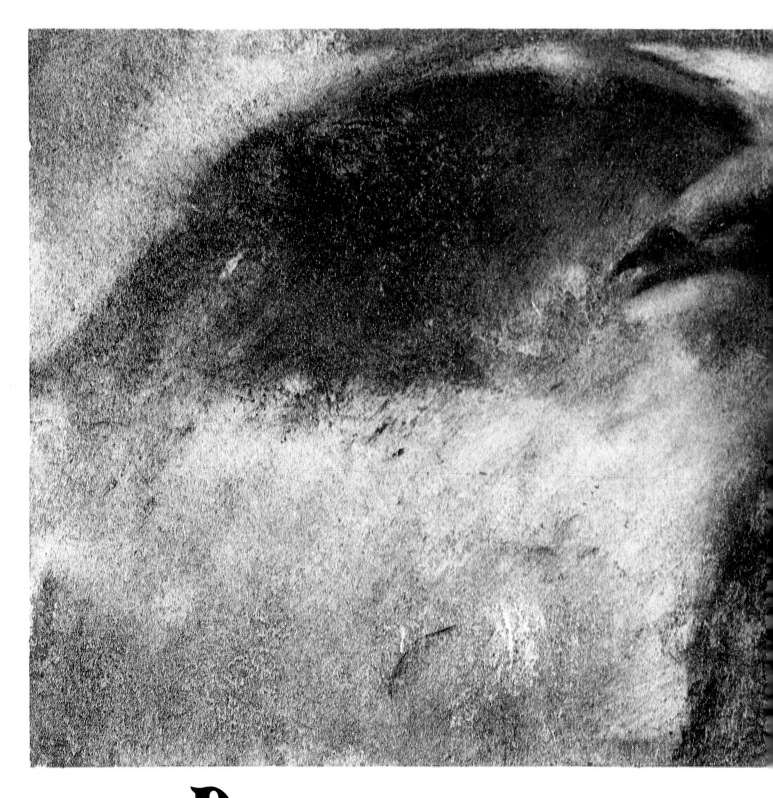

Dragons, ogres and monstrous birds guarded the slopes of the Shans' divine peak, Mount Meru, while sweet waters washed its shores. In certain seasons, the nymphs who made their home in the waters were drawn up onto beaches glistening with gold, silver and myriad gems to dance and play with one another, bathing in the hearts of lotus leaves and splashing in the shallows. Lusting after such beauty, the ogres left their posts to join the nymphs and immediately grew jealous of each other.

They fought, seeming to set the mountain ablaze with the flash of swords
and shields. Mistaking those flashing lights for a conflagration, the king of
the dragons sprayed the slopes with water from his mouth. The lord of the
great birds, thinking that the sacred mountain was flooded, beat the air
with his terrible wings to drive the water back. The winds that swirled
from his wing tips swept down the mountainside. And all these events
together made the hurricane.

Storms over Africa were the work of the god Kayura. When he wished to create a tempest, he tossed his flock of red-feathered birds down into the clouds. The storm birds flew like arrows loosed from heaven, their glittering plumage aflash with lightning. Thunder echoed in the flapping of their wings. Sometimes the lightning birds struck the earth and buried themselves in the ground, never to emerge again.

The Heaven Doctors of Natal—tribal wizards—knew how to catch the storm bird by luring it to them with a bowl of food. Once they had succeeded in capturing one of the storm god's feathered minions, they could conjure tempests in a clear blue sky themselves. Over a fire they rendered the bird's fat, and dipped the ends of their wands in it to summon a thunderstorm at will.

Three

The Dance of Life

The Primal Potter

There was a time before time when animals and humans lived together in greater intimacy. Those beasts tamed for use or cherished as pets sheltered under the same roofs as their owners and, in their own ways, worshipped the same gods. Even the hunters and the hunted treated one another with respect. No trap was ever set, nor knife unsheathed, without a word of grace to those creatures that gave up their lives to provide their two-legged brethren with food or fur.

Different species knew one another's habits, studied one another's virtues and vices, and—it was rumored—shared a now-forgotten tongue. Stories may have passed between them, for the animals had long memories. But even if they were not the source of all old narratives, they were often the subject of them.

Deep in the jungles of central India, remote from the trade routes, temples and cities, the ancient forest peoples preserved certain tales of great antiquity, which explained the world and its human and animal inhabitants. The nature implanted in each creature at its creation, and the outcome of its primeval encounters, governed forever the shape of its enduring relationships with humans and with other animals.

The earth and all things in it—so the old stories taught—were created by a supreme being called Singbonga. Once he had formed the mass of the earth and flung upon it the seeds of all plants and trees, he was ready to create the myriad tribes of living things.

He started with the ancestor of all horses. When that was dry, he turned to the making of people: Pressing and molding the moist clay, he formed the figures of men and women and left them to dry overnight. But the horse was afraid that if humans once appeared in the world, it would not be very long before they subdued the horse to their will and rode arrogantly upon its back. So, under cover of darkness, the horse trampled the soft clay of the figures into the damp earth, and in the morning Singbonga found that his work had been spoiled.

Setting to work again, he first made the figure of a dog, then a new set of people. This time he laid them out to dry so that they faced into the blowing wind. By nightfall, the dog was dry and the breath of the wind had blown into its nostrils, bringing it to life. But the clay forms of the people were still moist and soft, so Singbonga put the dog on guard to protect them.

During the night the horse came again, seeking to destroy its enemies. The dog leaped and snarled around the figures, and its barking drove the horse away, so that when Singbonga returned the next morning, the clay had dried undamaged. Then the spirit breathed life into it. And thus was the human race created, and the nature of its relationship with the horse and the dog established for all time to come ☆

A Tale of Many Tails

The braggart's bungle

Long before the days of Genghis Khan, the graceful horse and the oddly shaped camel lived side by side on the cold Mongolian plains. The nomad tribes told a story explaining how the two flat-footed beasts came to be so different.

The Creator, they said, made the Asiatic horse and gave it a neck that arched in an elegant curve. But Bertik Khan—who may have been a rival god or a demon—was jealous. He tried to create an even more beautiful animal. All thumbs, he formed the back with a hump in it, and put the neck on upside down. Whether Bertik Khan thought his creature handsome, no one knows, for he was never seen again.

In the old world, power took the shape of a pyramid with the monarch alone at the apex, his small band of wealthy land-owning nobles below him and the broad mass of commoners at the bottom of the structure. According to the ancient tales, this was the way of life not only in human society but also in the realm of the animals. The beasts and birds had their hierarchies and social classes, their laws and customs, their manners and mores, all of which disappeared from the face of the earth when the new two-legged species established its hegemony.

But before that conquest was complete, the lion presided as the king of the beasts. Slavic chroniclers said that a great lion ruled as Tsar of Russia in the days before humans. His vassals and courtiers were lesser creatures that bowed low before his wise brow, his awe-inspiring voice and his bone-cracking jaws. From the lattice-roofed throne room of his elegantly furnished summer residence to the tapestry-clad chambers of his Winter Palace, the lion lived in splendor worthy of his magnificence.

Yet his subjects did not covet the jewels that sparkled against his dark mane, the wine that filled his goblet or the company of servants that waited upon his roar. The elk of the mountain and the hare of the steppes alike desired only one thing that the lion possessed: a tail.

The lion's tail was supple and sinuous. It had a language all of its own, sometimes lashing with rage and at others curling with pleasure. It was extremely handsome, a thing of sleek fawn tipped with a black plume. And it was unique.

Early one spring, when the snow drifts that leaned heavily against the windows of the Winter Palace were just beginning to melt, the Lion Tsar summoned his retinue for the long journey to the Summer Palace. Carried in the imperial palanquin on the backs of his four-legged servants, he traveled the length of his land and saw how terribly his subjects suffered from their want of tails.

The Lion Tsar observed that when the bitter night wind blew into every burrow, the silver fox could not bury its nose in a thick, warm tail. As the days grew milder, he watched flies bite greedily into the flesh of the horse that was helpless to fend off their attacks. On the steppes he noticed that the antelope had no tail to flutter like a signal flag when the leopard approached. When the antelope fled, the leopard had no tail to twitch in anger. In the mountains, he heard the red wolf howl with sorrow because it had no tail to wag with joy.

Long before the onion-domed spires of the Summer Palace appeared beyond the forest, the Lion Tsar had resolved to end this misery. Arriving in his throne room, he sprang from the palanquin. He was determined to give every one of his subjects the benefit of a tail.

To achieve this, he called a council of his most learned scholars, wiliest ministers and most powerful wizards. All the masters of the highest arts and deepest sciences were summoned to contemplate where in the world tails might be obtained. Even seekers of truth normally dismissed as mountebanks were consulted: wild-eyed eccentrics that sought to transmute base metals into gold, others that played curious games with numbers or boiled up unspeakably noxious substances in carboys and retorts.

To all he posed the same question: Where could tails be found for the creatures that, unlike the lion, were unfortunate enough to have been born without one? Did they grow like rushes on the riverbanks? Could they be knitted, spun or woven? Were they mined beneath the earth like rubies, tin or silver? Could they be made by artisans laboring at long benches in windowless workshops?

The convocation to ponder the problem must have been held in secret, for no record of its deliberations survived. But the Tsar and his company of advisers somehow came up with an answer. A few weeks later, in the dead of night, a number of heavy coffers, jars and baskets were carried clandestinely into the palace and presented to the lion.

The monarch opened each container and examined the multicolored tails within, running his paw along their textured lengths, hefting them to judge their weight and substance. The lion was satisfied. Roaring up to the birds that wheeled under the vaulted glass of the ceiling, he commanded them to fly with a message for all his subjects. Every beast that came at once to the palace would receive the tail it desired. In a rush of wings the birds were gone, soaring across the kingdom and singing the words of the Lion Tsar.

The elk and the bull lifted their heads from the spring grass and listened to the message. Over the mewling of their young the badger, the fox and the sable heard the song. The lynx and the leopard left the hunt to obey, followed by the hare and the reindeer they had chased.

Soon the birds had called every beast in the land but one. No matter how high they soared or how low they swooped, they could not find the bear.

They looked by the river where it usually fished. They searched in the forest where the sweetest berries grew. Despairing at last, they were heading back toward the Summer Palace when a snore rumbled up from the very earth itself.

The birds wheeled and dived until they found the mouth of a rocky den. Inside was a fat ball of fur snorting steamy breaths and growling softly to itself—the bear, asleep, winter still in its bones.

The birds flew at it, pecking its paws and pulling its fur. As it stretched and scratched, they told it of the Tsar's gift. The bear shook itself awake, rolled to its feet and shambled out into the sunshine.

It found the road to the palace muddy with the tracks of all the beasts that had

In the beginning only the lion, Tsar of all the animals,
possessed a tail of his own. But in an act of royal largesse, he summoned his subjects to his
Summer Palace and provided tails for all.

set out before it. Following along, it sniffed the spring air which smelled of rain, grass, flowers and honey. The bear stopped. It had eaten nothing during all the months of its winter sleep and the road ahead was long.

The honey scent drew it from the road to a hollow lime tree in the forest. At the top of the tree hung a beehive, heavy and dripping, sending down streams of sweet nectar. The bear feasted. Honey ran down its chin and dribbled onto its fur. Soon the hollow tree was empty and the bear was full. Looking down at its sticky coat, it decided that it must bathe before appearing at the Tsar's court. It splashed the honey away in the river, then lay down to dry itself on the sunny bank. Before long, it was asleep.

In the Summer Palace, meanwhile, a line of beasts stretched from the foot of the throne, around the samovars and under the icons all the way to the gate. At the head of the line was the fox. It had arrived at the court first and begged to be allowed to choose first among the tails. The Lion Tsar nodded his assent, opening one of the many jeweled caskets that lay before him.

The fox held up first one tail, then another. The crowd of animals gasped as they saw curly tails, bushy tails, bristly tails, silky tails, tails in gold and russet, cream, black, white and silver. The fox chose the finest one, a luxurious brush, but there were many magnificent tails left. The horse took a long, slapping one to keep off the flies; the squirrel found a soft, thick one to keep out the cold.

The sun rose high and sank low as each beast made its choice. The Lion Tsar's eyes were heavy by the time the hare, last in the line, bowed low before him. The Tsar handed the final casket down to the hare. It was empty. All the tails had been taken. Nevertheless, the hare scraped up a bit of fluff that clung to the bottom of the coffer and bounded away, content.

As the night grew cool, the bear awoke by the river. It snuffed and sputtered and lumbered toward the palace. It arrived by moonlight to find the windows dark and the gate locked. As it turned away with a bitter snarl it heard a sound coming from the shadows. Peering into the darkness, the bear could just make out the badger, twisting and turning to admire its new striped tail.

The bear admired it, too, although it thought it looked too grand and rather gaudy on such a small animal. It asked the badger to give it the tail. The badger barely looked up from its new finery. This was an act of profound discourtesy, for the bear by virtue of its size and strength far outranked the lowly badger in the hierarchy of forest creatures.

With a growl, the bear planted a heavy paw on the tip of the badger's tail. The badger wrenched free, taking most of its tail with it. Only a tuft of fur, caught in the bear's claws, was left behind.

The bear prodded and poked the ragged clump, patting it into a tail—a very small tail. Then, with a sigh, the great beast fixed it onto the spot where it remained forever after ☆

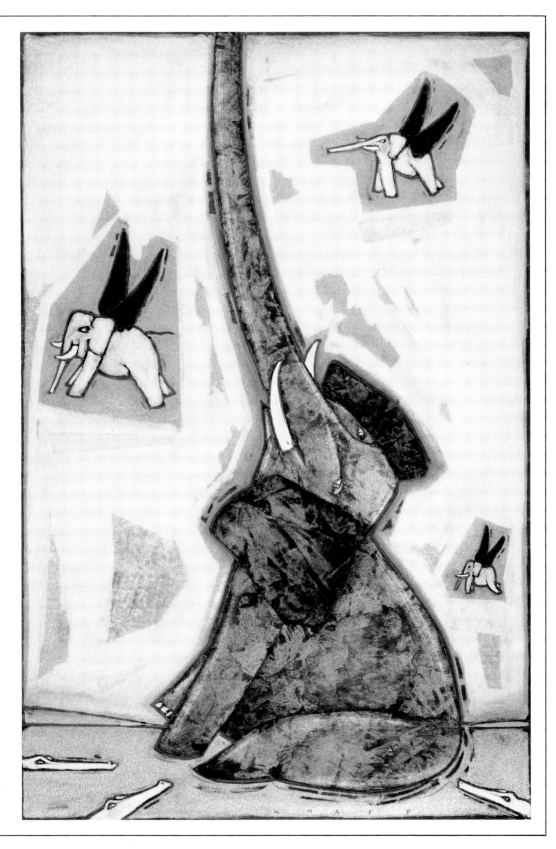

How the elephant lost its wings

In the beginning, said the wise men of India, elephants could fly. But through an unhappy accident, they lost their wings and became earth-bound. It happened this way:

An elephant flying over the hot countryside spotted a particularly cool and tempting pond. Descending rapidly from a great height, it crash-landed on the reedy bank. The impact shook the ground and muddied the water, and disturbed an enormous crocodile from its sleep. While the elephant noisily slaked its thirst, the crocodile attacked.

The elephant tried to fly away, but one of its wings was clamped in the crocodile's jaws. It is said that the elephant and the crocodile then fought for twelve years and thirteen ages before the elephant, both wings now gone and its hide slashed to ribbons, called upon the great god Bhagavan for help.

Seeing that the elephant was nearly drowned, Bhagavan reached out of heaven, grasped the beast by the nose with one hand and pulled. In those days, the elephant's nose was smaller. As Bhagavan pulled, the nose stretched, but the elephant did not rise. So Bhagavan took hold of the elephant's ears with both hands and, giving a mighty tug that flattened and widened them, yanked the great weight from the pond.

The elephant soon got used to its new shape and a life without wings. The people of that part of India were particularly pleased, for they were tired of low-flying elephants crashing into their cowsheds.

A Red Badge of Courage

The stories the ancients told of magical and heroic struggles, mythic beasts and talking animals were not spun to entertain. They were woven of people's darkest fantasies and deepest fears. This web of words and images set up a potent magic that could keep out the evil forces that figured in the narrative and threatened the life of the tribe. And throughout the north, no creature was more dreaded than the giant bear. Of all the beasts, it was this one that seemed most human and therefore most dangerous. The female bear cared for its young as a human mother would, and stalked its prey just as humans did. Walking on two legs, the bear was more than a match for even the most skilled of hunters.

The great white ice bear haunted the dreams of the Eskimos. The hunt was more than a battle between hunter and animal, it was a struggle between man and mortality in a forbidding landscape where death was always on the prowl.

It was not only the weakest members of any group—the old, the sick, the injured—who were at risk. Even the strongest hunters came close to death every day. It was only through the potent magic they gathered from the spirits of the creatures they had killed that the hunters were able to survive.

Against this background of relentless snow and cold, a story was told of a boy and his father and their struggle against the most dangerous forces of the north. Into this tale were woven the elements the Eskimos most feared: separation from the family, winter, illness and above all, the bear. Its telling constituted a potent weapon against these dangers.

It was said that the boy and his father traveled across the land, alone, the only human beings in a frozen world. How they came to be on their own, the story does not tell. They may have been sent as a punishment for a crime against the great white bear itself. Perhaps the father had broken one of the taboos that allow men to hunt and kill a beast so much greater than themselves. He may have failed to make the proper speech of apology to the slain bear or neglected to offer the dead creature water or, worst of all, he may have forgotten to dedicate the skull of the first bear he killed to the sea. Whatever caused them to wander, they were far from the protection of their group when the story began.

The boy did not mind much. To him, his father was a great hunter, storing up immense stocks of seals, caribou and walrus in cairns in preparation for the coming winter. It did not matter to the boy that his father was silent and grave, for that was just his way. The child spent his days snaring small animals and playing at being a great hunter himself. It was also his job to gather driftwood and the mosses that had burgeoned in the short Arctic summer; these would fuel their fires during the winter months.

On his wanderings, the boy came upon a bird, smaller and quite different from any he had ever seen before. Stealthily and slowly, he took up his lance and crept cautiously toward the bird. Small though it was, there would surely be some meat

The great ice bear, bringer of death, hovered in the shadows of an endless winter night. It sought—and often found—vengeance against the hunters who plagued it.

on its tiny bones. Instead of flying away at his approach, the bird turned toward the boy, cocking its brown head to one side and fixing him with a bright, almost welcoming stare. The boy lowered his lance in respect; the bird must have a very powerful spirit to be able to look into the face of death.

From that day on, the boy often saw the bird in his travels. Whenever he did, he stopped to speak to it, never forgetting to give the bird an offering of food from his pouch. He learned that the bird liked the insects and slugs that hid under the lichen-covered rocks, and after a time the bird was bold enough to feed from the boy's hand. With the magic of the strange creature on his side, the boy hoped fervently that he would become a great hunter.

When the darkness of winter settled in, the boy and his father retreated to the shelter of their snow hut. They ate the meat and fish they had been storing up and passed the hours by a small but bright fire of mosses, driftwood and twigs that the boy had so painstakingly collected. The igloo was well insulated by the snows that piled on top of it, and the boy and his father were kept warm by three layers of furs.

At the beginning of the winter, the man still went off to the kill, leaving the boy to tend the fire in their home. But at last the days drew in and night once again ruled the world.

This was the season of the ice bear. As the man and the boy huddled together by their flickering fire, they could hear breathing outside the igloo. The great beast was at hand. In places where people lived in groups, as they were meant to do, bears did not dare to come so close. They were afraid of men's lances and of the sharp teeth of their dogs. A bear had no fear, however, of two lone and defenseless humans, one a mere boy.

So the bear snuffled around the igloo, looking for an opening, savoring the tantalizing smell of the people inside. From time to time, the bear peered into the narrow entrance of the igloo, and its close-set eyes would glint for a moment beyond the firelight. Then, afraid of the fire, the animal retreated.

The boy and his father said nothing of what they felt, but they were both terrified of the beast that stalked them. To speak of their fear would only add to the bear's power. But they knew that if they told stories about great bear hunts and the many ways in which bears had been killed by men, it was possible to keep the creature at bay.

The man began to tell of all the hunts he had ever seen, of all those he had ever heard of, of those he had been on himself. As the bear waited outside, its belly rumbling, the man spoke of bears surrounded by snarling packs of wolf-dogs whose teeth nipped their thickly furred flanks, while a man raised his spear ready to strike deep into the bear's heart.

Wide-eyed the boy listened as his father described the bravery of hunters and the mortality of bears. Tale followed tale through the long hours of darkness as man and boy huddled together for warmth, feeding scraps of precious fuel to the flames.

At last, the man's voice grew hoarse, his brow became feverish and his eyes glimmered with delirium. He fell into a deep sleep, and his sobbing son could not wake him. Struggling bravely against his tears, the boy took up the thread of the tales. His thin childish tones murmured on for long hours, but now his only audience was the bear outside.

With trembling fingers, the boy added precious twigs to the fire one by one. The bear's eyes glowed in the doorway, full of anticipation. Desperately fighting sleep, the boy spoke on, weaving bloody tales which always ended in the death of a marauding bear at the hands of the courageous hunter. At last he could speak no more, and he fell into a troubled sleep by his father's side.

As the fire gradually weakened and died down, the bear grew bold. First it thrust one of its massive paws into the igloo, beating down the embers until there was nothing left. Then it stretched its paw in further but still could not reach the two sleepers. With a mighty thrust of its shoulders, the bear tried to crash through the narrow doorway of the igloo, but the hard-packed snow would not give. Disgusted, the bear wandered away to seek the warmth of its own lair, confident that its first meal of spring would be the frozen flesh of the man and boy it had stalked so patiently.

The wind howled outside the igloo, master of the empty Arctic landscape. On its chilling breath it brought the small brown bird the boy had befriended long ago in the days of light. The story does not say how the bird knew of the boy's plight or how it came to be in the Arctic out of its season.

In any event, the little creature saved the boy's life. Scratching in the embers, the bird found a spark and, carefully fanning it with its wings, brought it to life and brilliance. The bird fanned on, coaxing the dead embers to burn once again. The flames rose up, scorching the bird's breast and making the boy stir in his sleep. As the boy's eyes opened, he saw the bird, its breast now red with the heat, flying from the tent.

From that day forward, robins had red breasts. The glowing feathers were passed on from one winged generation to the next, as a badge of honor bestowed upon the robin to mark the gift the bird once gave to humankind ☆

A feathered Heraldry

In days long past, when kings and princes wore gorgeous costumes to dazzle and impress their followers, it was believed that the flamboyant plumage displayed by certain birds was a magical gift—a token of gratitude, perhaps, or an emblem of heroic deeds.

Arab chroniclers declared that King Solomon gave the hoopoe its golden crest. It was written that Solomon, Israel's wise and powerful king and the master of all magicians, was one day soaring on a flying carpet high above his lands; the sun was blazing down from a cloudless sky. With no protection from the intense heat, he was near collapse when a flock of hoopoes flew to his rescue and formed a canopy above him to shield him from the sun.

When the journey ended and the carpet came to rest, Solomon allowed the hoopoes to name their own reward. The proud birds chose crowns of gold, as if they were kings themselves. But this act of presumption led them to disaster: Robbers hunted the hoopoes and stole their crowns. Chastened, the birds returned to Solomon and begged for his indulgence; and the mighty sovereign gave them new crowns of golden feathers. It was not only exotic species like the hoopoe whose attributes were imparted by magic. The blackbird, according to a French legend, acquired its color in a foray to the bowels of the earth in search of the realm of Mammon, Prince of Wealth. Before setting out, the bird had been warned not to touch the heaps of glittering treasure that lined the subterranean passageways.

But the creature could not resist temptation, and plunged its beak into a mound of gold. Suddenly a demon made of smoke and fire burst from the shadows and chased the bird out of the cave. Ever after, its beak was yellow, its feathers ebony, its startled shriek a souvenir of that moment of terror.

The blackbird was not the only one to grapple with a denizen of the underworld. The folk poets of Latvia told a tale that explained the origins of the woodpecker's blood-red cap. The peasant bards, whose people carved their living from the bleak but fertile plains bordering the Baltic Sea, saw the woodpecker as an innocent victim in a plowing competition between God and the Devil.

Each had to plow a field of exactly the same size, but God started the match at a

Arab scribes said that the hoopoe once wore a crown of real gold as a gift from Solomon.

serious disadvantage: While the Devil had a team of powerful horses, God's plow was drawn by just one tiny woodpecker. By evening the Devil had finished half his field, but the woodpecker had hardly completed a furrow. A desperate measure was called for if God was to win.

That night, while the Devil slept, God crept into the next field and borrowed the horses. Throughout the hours of darkness he plowed steadily on, but just before dawn he took the horses back. When the Devil returned, he was so astonished at God's progress that he demanded they exchange plow teams. God agreed, of course, but as he drove the horses to victory the Devil saw he had been duped and took revenge on the poor woodpecker. Smashing the bird's head against the plowshare, he left it battered and bloodstained for all eternity.

In partial compensation, perhaps, the bird was endowed with the ability to predict the weather. To the farmers of old Europe, the sound of the woodpecker tapping in the trees was a sure indication that rain was on the way.

According to the poets of ancient India, the peacock, too, could forecast the weather, uttering its harsh cry when a storm was due. And like the woodpecker, this bird owed its appearance to divine action. Originally it was a drab creature. Then Indra, god of war, fleeing an attack by Ravana, the many-headed demon, concealed himself inside the plumage of a peacock. The ruse was so successful that the god, as a mark of gratitude, decorated the bird's tail with a thousand golden eyes.

Strutting proudly with its tail fanned and erect, the peacock became a symbol of royal splendor. The monarchs of the ancient Orient sent peacocks in tribute to King Solomon's court, where they easily outshone the hoopoes. But the peacocks had cause for humility, for their magnificent bodies rested on clumsy feet. It was said that whenever the peacock glanced down it screamed in horror at those ugly talons and its tail sank in despair—the very image of wounded vanity ☆

The peacock's delight in displaying its gorgeous plumage made it a symbol of vanity; yet nature gave the bird its own comeuppance by underpinning this glory with a pair of ugly feet.

The fox and the fishes

In the morning of the world, says an old Jewish legend, the vast seas were empty except for the huge bulk of the monster Leviathan, lurking at the bottom of the ocean. He was a king without subjects until the Angel of Death was sent to populate the seas by drowning one member of every species of land creature and transforming it into a fish.

The fox determined that he would outsmart the Angel of Death and cheat the Leviathan. As he sat on a bank beside the sea, contemplating his watery future and wondering how he could escape it, his reflection gave him his cue just as the shadow of Death fell upon him.

Instantly, the fox burst into tears and loud lamentations.

"Why do you cry, Fox?" asked the Angel, impatient to get on with his work.

"I am mourning my friend," said the fox, sobbing. "As your shadow passed over him, he threw himself into the sea in his haste to join the Leviathan's legions. There he is now." The fox waved sadly at the creature in the water who waved sadly back at him.

"Good, good," said the Angel, and flew away.

All went well for the fox until a year later when his deception was discovered by Leviathan himself. During the counting of the fish, he realized that there was no fox fish among them. Displeased, Leviathan lashed his dragon-tail through the waters, demanding to know why. The timid parrot fish told how the fox had tricked the Angel of Death.

"Bring me the fox alive," the Leviathan commanded the catfish. "I wish to eat his heart and thereby gain his cleverness. Tell him that I am dying and wish to make him King of the Fish in my place."

The catfish soon found the fox, and told him Leviathan's story. Proud of the honor, the fox hurried onto the catfish's back.

On the long journey, the fox had

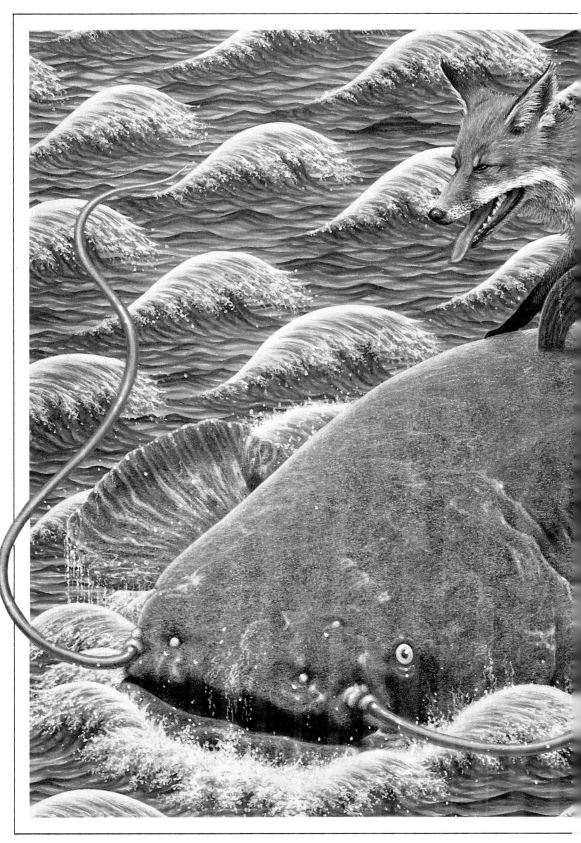

time to reflect and wondered if he had not been tricked. "O Catfish, now that I can't escape, tell me what the real purpose of this trip is," said the fox. The catfish revealed the Leviathan's plan with great satisfaction. Fox was not so clever after all, he thought.

"My heart!" cried the fox. "He wants to eat my heart! Now you are in trouble for I haven't got it with me. Why didn't you tell me while there was still time? Didn't you know that we foxes never carry our precious hearts with us? It is back home, safe in my burrow."

The fox suggested they return to shore to retrieve the heart. When they reached land, however, the fox jumped off and scampered away, jeering at the catfish's stupidity. The catfish hid beneath the bank, where he remained, afraid to face the wrath of the Leviathan. The fox has never returned to the shore, which is why to this day there are no fox fish in the sea.

A muddy Metamorphosis

The ancients knew not only of birds that had been honored for their kind deeds with crowns and feathers of brilliant colors, but also of men dishonored by their actions, who were transformed into loathsome beasts. Such tales served as a warning to the wicked.

In her enforced wanderings around the Greek isles, the goddess Leto, mother of Artemis and Apollo, met many mortals. Most of them showed her the kindness that was due to all travelers. In Greece, where any stranger might be a god in disguise, hospitality was a duty that no prudent person would shirk: Those who withheld it might pay the price of death.

Leto wandered because she had been cursed by the jealous Hera, queen of the gods. Leto's crime was that Zeus, Hera's husband, had loved her and that she had borne his children. Now she was doomed to roam the earth, her divine children in her arms, never finding a place to rear them. The ancients tell how on these travels she came to mete out an extraordinary punishment herself.

Having walked without resting for many days, Leto came to the crest of a hill and saw, spread out below her, a perfect valley with a willow-fringed pond at the bottom. Leto's throat was dry, her lips parched and cracked, her skin covered in a film of dust. She longed to drink from the pond's sweet, cool waters and hurried toward them.

Around the pond, some country folk were cutting willow branches. The men's rough shouts and tasteless jokes, made louder and more raucous by the strong wine which they had been drinking in preference to the pond's sweet water, rang out in the shaded glen.

Leto barely noticed them. Putting the infant gods down to play on the grassy bank, Leto knelt to drink, but before her lips could touch the water, a rude voice called out, accusing her of trespass and ordering her to stop.

Leto straightened up wearily before replying. All around her, insults and jeers echoed. The goddess begged her tormentors to allow her to slake her thirst, although there was no need to humble herself in this way. She knew that nature had been created for all, be they divine or human, to enjoy, not for any mortal to control. Still, her words were greeted with oafish cries and coarse remarks. Leto blushed with fury but she restrained her anger and again asked to be allowed, for the sake of her children, to drink.

The men clearly delighted in wielding their brutish strength against a lone, seemingly defenseless woman. Laughing still, they jumped into the limpid waters of the pond and started to stir up the mud from the bottom, making the waters filthy and unpotable.

But Leto had reached the end of her patience. Raising her arms to call upon the powers of Olympus, she sentenced her persecutors to live for an eternity as pond creatures, beings who delighted in mud and filth, whose throats swelled with uncouth croakings, whose mouths were stretched by their own vulgarity, whose bellies were pale and bulbous. In this way, frogs were created☆

Seeking to quench her thirst in a pond, the goddess Leto was abused by hostile peasants who muddied the water before she could drink. In her wrath she punished them for their cruel inhospitality by turning them into frogs.

115

A Glorious Genesis

To the early sages, the world of nature was a living textbook of moral tales and edifying allegories. Leto's mottled frogs, condemned to a life in clammy places, reminded mortals that the gods were swift to punish acts of unkindness. But more beautiful creatures were greeted as promises of hope and redemption.

Never were such portents needed more than they were in one of the darkest ages of ancient China, at a time when an emperor of exceptional greed and cruelty sat on the throne. He presided over a court riddled with corruption, where bribery was the common currency and flattery the only language spoken.

The monarch remained aloof from his subjects, in the palace that astrologers identified as the precise center of the universe. He passed his days counting the gold, receiving the reports of spies and condemning those who displeased him to slow and terrible deaths.

It was left to his prime minister, a venerable mandarin, to oversee the management of the empire and ensure that the imperial will was carried out. But one day, for reasons the chroniclers do not explain, the minister became sickened by the corruption around him. Pleading ill health and the increasing feebleness of age, he obtained the ruler's grudging permission to retire from the court. There would be no shortage of power-hungry younger men to take his place.

Emerging, for the last time, from the palace, he ordered his palanquin-bearers to carry him to a monastery some distance from the capital. Here, the mandarin planned to spend the remainder of his days in prayer and penance.

But soon after the smoke and clangor of the city were left behind, he commanded his servants to stop at the edge of a woodland. Emerging from his curtained litter, the minister told them to return for him at sunset. He stood for a moment breathing in the sweet, clean air. Then he sank down and wept for his sins.

Afterward, he sat quietly, listening to the bird song. So still was he that bees lit upon his brilliantly colored clothing, mistaking it for a bed of wild flowers. On an impulse, he drew out his dagger and cut a piece of silk from his robe. He placed it on the ground to see if the bees would be fooled again. The bright scrap rose and fluttered, as if lifted by a breeze. But there was no breeze. The mandarin looked harder: The fragment seemed to be flying of its own free will. Amazed, he cut another snippet; it too took wing.

His dagger glinting in the afternoon light, he sent more embroidered scraps into the air. Soon the glade was alive with delicate creatures, flashing their colors and patterns as they danced.

When at the day's end his servants came for him, they found him sitting on the ground, his magnificent robe in tatters. But the tears running down his cheeks were tears of joy. Miraculously, he had at last achieved something beautiful and good. He had given the world its first butterflies ☆

The Garden of the Gods

For the ancient Greeks, the plants of field and garden had supernatural tales to tell. They were living proofs of the Greek gods' stormy and unpredictable ways— with one another as well as with the men and women they encountered.

The humble violet, for example, was one of the results of a flirtation Zeus, lord of the heavens, had with a pretty nymph called Io, the daughter of the river-god. When his long-suffering wife, Hera, became suspicious, Zeus quickly turned the nymph into a white cow and proclaimed his innocence. This did not fool Hera, who sent a gadfly to torment Io. Zeus, touched by some slight remorse, created violets as a delicacy that Io might nibble in her new bovine form—a poor return for the nymph's affections.

Even Zeus's rare bouts of marital fidelity sometimes served a botanical purpose. The crocus, it was said, sprang into life on a mountainside made miraculously warm in a spot where Zeus and Hera had made tempestuous love. Another story, though, insisted that the crocus began as a handsome youth in the garden of the goddess of flowers. Unwisely, Crocus spurned the love of one of her nymphs: At a gesture from the irascible goddess, he sank down into the earth and then re-emerged as a flower, gaining immortal beauty at the cost of his human form.

Many a plant was evidence of the risk mortals ran if they attracted unwanted notice from a god. The fair maiden Rhodanthe, for instance, was once seized by a pack of unwanted suitors and carried to a pedestal reserved for the moon goddess, Artemis. The girl protested furiously, but anger only served to enhance her beauty enough to catch the eye of Artemis' brother, Apollo, god of the sun. Outraged by the insult to his sister, Apollo turned Rhodanthe into the first rose; and as a relic of her fierce pride, he gave the rose its thorns.

Another legend attributed the rose's thorns—as well as its color—to little Eros, the child-god of love, whose irresponsible actions often had serious consequences. Once, all roses were white, until Eros upset a cup of Zeus's potent red nectar and dyed the flowers forever.

The creation of the rose's thorns involved Aphrodite, Eros's mother, too. On an earthly visit, Eros chose to kiss the petals of a rose, his favorite flower. But inside the blossom there lurked a bee, which promptly stung the boy upon his lip. His pitiful cries of pain drew Aphrodite to him; in an attempt to comfort her son, she drew the stings from all the bees in the rosebush and used them to stud the flower's stems with thorns.

Eros also had a part in the making of the graceful laurel. Once, with his usual malice, he pierced Apollo's heart with a love-arrow. For the first time, the sun-god felt the pangs of love, which sent him panting after a wood nymph named Daphne. But Eros had struck out at Daphne, too, this time with a leaden arrow that drove all thoughts of love from her heart and made her yearn for a life of maidenhood. Thus, when Apollo made advances to her, she was horrified and fled from him through the woods until she came to a river barring her path. There, with the impassioned god's hot breath upon her, she prayed for help to the river. At once, her legs took root, her skin turned to bark and Apollo was left with nothing but a lovely tree whose leaves thereafter he held sacred.

Even without a spur from Eros, Apollo was powerfully attracted to beauty—in either sex. Once, he was greatly taken by a handsome youth called Hyacinthus, and for a time the god and the mortal became inseparable. One day, tired of their usual sport of hunting in the forest, they stripped off their clothes and oiled their bodies in the manner of Greek athletes and engaged in a friendly discus-throwing competition. The sun-god made the first throw—which turned out to be the last. Either by accident or, as some claimed, with the help of a jealous nudge from the West Wind, the returning discus struck Hyacinthus full on the face and he sank, dying, to the ground.

Apollo was distraught with guilt and grief, but even he could do nothing to save his lover's life. Instead, he changed him into a sweet-smelling flower that would always bear the sign of the god's great sorrow. The boy's red blood was transformed into purple blossoms.

As a rule, the amorous intentions of the deities caused mortals nothing but trouble, though without them the world of plants would have been much smaller. The fir tree, for example, was the result of the passion of the mother-goddess Cybele for one of her own priests, a young man called Attis.

The goddess was relentless in her pursuit of the unwilling youth. Eventually, unhinged by the ferocious jealousy of his divine mistress, Attis took drastic action to escape her insatiable desire: With the help of a sharp stone he castrated himself. Even that was not enough, and Attis was on the point of killing himself when a kindly Zeus intervened, transforming him into a fine, upstanding fir.

Sexual dalliance with a god or goddess was dangerous enough in its own right, but most of the deities had a wife, husband or lover whose wrath had also to be reckoned with. Mint was one plant that owed its beginnings to a jealous spouse. Hades, god of the underworld, was something of a stay-at-home by divine standards, and his gloomy kingdom of the dead claimed most of his attention.

On a rare emergence into the upper earth, however, the god caught sight of an exquisite nymph called Minthe. Love thundered in his dark old heart, but his wife Persephone was far from pleased. She herself was not overly enamored of her saturnine husband; in fact Hades had been forced to kidnap her to bring her to the bridal bed. But she refused to accept her husband's philandering with such a young and beautiful girl, so she changed

Eros, capricious messenger of love, was attacked by a bee concealed within a rose. Appealing to his mother Aphrodite to punish the flower that had caused him pain, he was satisfied when she attached the insect's stinger to the flower's stem. Forever after, all roses were armed with thorns.

her rival into an herb that could attract a man with nothing more than its cool, refreshing flavor.

Sometimes, a mortal could achieve translation into planthood more or less unaided, or at least without the intervention of the gods. Such was the lot of the celebrated Narcissus. He was not, perhaps, an ordinary mortal, since he had been born to a river nymph who had been raped by a river; but in the normal course of events he would have died like any other man. Narcissus, though, was extraordinarily beautiful—as a baby, as a boy and finally as a young man.

Wherever he went he left a trail of broken hearts, both male and female, behind him, for he scornfully disdained the embraces of anyone less perfectly formed than himself. Even the nymph Echo, condemned for a previous misdemeanor to lose all powers of speech except for a sad repetition of the last few words that were spoken to her, languished with hopeless desire for the haughty youth. When Narcissus sneered, "I would die before I would let you touch me," poor Echo could only answer, "Touch me!" And when the proud boy cried to the nymph, "Never could I love you!" all she could do was to sob out quietly, "I love you."

But one day Narcissus found his perfect match. Gazing absently into a still, forest pool, he found himself face to face with just the sort of gorgeous boy his heart had yearned for. He fell helplessly in love with his own reflection. For hours, then days, he stared at his beloved until his strength, his beauty and at last his life had gone. But when his companions came to give his corpse its ritual burning, they found no body. Instead, there was only the white flower by which Narcissus was forever remembered.

Thus many a man or woman left behind a new plant as a memorial. Sometimes, the transformation was granted as a kind of reward. The Athenians told just such a tale from the time of the Trojan War. Along with many other women of Athens, the Princess Phyllis waved goodbye to her lover at the onset of the war; but when the fleet returned home after ten years, the young man's ship was not with it.

Dawn after dawn the princess took her place by the shore, peering out across the blue Aegean, until at last she died of grief. Athena, goddess of Athens as well as wisdom, was so impressed by Phyllis' fidelity that she turned the princess into an almond tree. Her lover arrived the day after her death, and at his touch the leafless almond blossomed, as almonds continued to do ever after.

Athena featured, too, in the rare creation of a tree without first having to sacrifice some unfortunate mortal. Indeed, it was the means by which she earned Athenian favor. She and the sea-god Poseidon were once rivals for the loyalty of the people of Athens, until they agreed on an unusually peaceful way of settling the matter. Each gave the city a great gift; the citizens were to choose which was the greater. Poseidon produced nothing less than the first horse, but Athena created the olive and won hands down ☆

Heaven's homesick exiles

In places as far apart as Lithuania and Tierra del Fuego, people once viewed the sun and moon as husband and wife, and the stars as their children. According to some stories, their family life was almost exactly like that of human beings.

Father sun was a distant figure; some even said that he and the moon had quarreled, which is why they lived apart. Mother moon raised the children alone, teaching them how to shine and polish the lamps they held up in the sky at night. But the stars, like real children, often rebelled as they grew older. One night, some of them decided not to take up their lamps. Instead, they turned somersaults and sang taunting verses, refusing to listen to their mother when she ordered them to return to their chores.

The moon, however, was not a softhearted human mother, and she meted out heavy punishment. With barely a good-bye, she sent a wind to sweep them out of the heavens.

Dropped onto the cold earth, the stars begged to be taken back. Their pleas went unheeded until the next morning, when the sun learned what had happened. He explained that stars who fell to earth could never return, but he comforted his crying children by promising to let them brighten the earth as they had brightened the heavens.

He changed them into dandelions, who turn their shining faces upward to watch their father's daily journey. But there are always a few in every field who turn into puffballs, trying in vain to fly back to their mother on the wind that brought them.

Bequest of a Golden Stranger

In early days, humankind was compelled to kill or forage for all its food. Lives were short, hunger inevitable, and the harsh winters fatal to the weakest members of any tribe. Then the earth yielded up the secrets of planting and reaping.

These newly revealed mysteries were treated solemnly. The sowing of seed was a sacred marriage. Plowing and harvesting were acts of necessary violence upon the body of Mother Earth: As any midwife would affirm, there could be no life without the shedding of blood.

The time when this great knowledge was conferred upon the Iroquois nation remained engraved upon the memories of their storytellers. They spoke of a remarkable encounter in the wilderness, and of an ordinary young man who was given an extraordinary gift.

Always the young people of the tribes marked their passage into adulthood by withdrawing to the forest for a period of solitary fasting, prayer and dreaming. The hero of this tale, like all his peers, was sent out from his village as custom decreed. His mother marked his face with charcoal in the ritually prescribed manner, so the spirits would know he was a supplicant and take pity on him. Then she led him up into the hills to a lonely place far beyond the smell of the people's cooking fires. She cut branches and covered them with bark to make a shelter. Finally she left him there alone, without food, to wait and pray for a vision.

For three days he sat cross-legged on the ground or wandered among the trees chanting the spirit-bringing song. For three nights he lay awake inside the wigwam, listening to the sounds outside, trusting that his supernatural guardians would not allow a bear to maul him or a wolf to attack.

The boy waited, schooling himself in patience, for a spirit guide to bring some gift or message that would dictate the course of his adult life. He and his friends had often spoken of the powers they hoped to receive from the unseen guardians. Some wished for skill as hunters or healers; others aspired to longevity, to wisdom or simply to good luck.

Light-headed with hunger, he became suffused with a feeling of love for his people. Exhaustion gave way to ecstasy. He decided that, instead of asking for some private gain or personal glory, he would pray to be allowed to bring a lasting benefit to his tribe. But another day passed and no spirit came to him.

In the middle of that night, he awoke with a sharp sense that someone was watching him. He turned on his sleeping mat and saw no one. But still he felt a powerful presence. He rose, emerged from his shelter and encountered an extraordinary figure. Standing outside the wigwam in the moonlight was a tall man, heavily muscled, wearing a green robe and a headdress of golden feathers. The young initiate felt neither fear nor puzzlement. He had been taught to expect strange things in the forest.

The boy rose slowly, for hunger had weakened him. He greeted his visitor according to the custom of his people and apologized for the lack of food or drink to

In accordance with the customs of his people, a young Iroquois marked his coming of age by a period of fasting, prayer and meditation in the forest. During his retreat, he encountered a stranger clothed in green and crowned with golden feathers, who challenged him to a test of strength.

With straining sinews and flashing limbs, the youth and his visitor engaged
in a furious wrestling match. Although he was faint
from fasting, the boy's power surged up to make him a more-than-equal opponent.

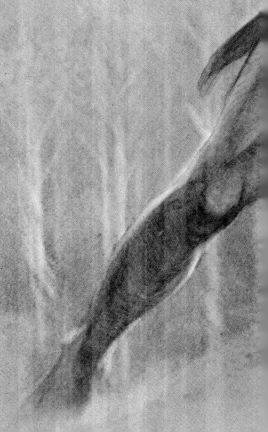

offer him. They conversed for a while, as any amiable strangers might, about the weather, the animals, the prospects for the next season's hunting. The man questioned the boy about his home village: Had any newborn children survived the last hard winter? Was the hunting good? Then the stranger challenged the boy to a friendly wrestling bout.

The pair stripped down to their deerskin breechclouts and began the game. As they grappled, the boy's hunger evaporated. He felt full of sap and strength. But the stranger was possessed of a power that seemed more than human, and the Iroquois required every ounce of stamina he could muster to stand his ground. As he grappled and thrust and twisted, he felt himself drawn into a trance. Night turned into morning and the red glow of the dawn washed their bodies with a coppery sheen. Unaware of passing time, they moved through the measures of a dance of combat, their pounding hearts beating out the rhythm. Only when the sun sank low in the heavens did they stop. Breathless from their exertions, they lay side by side on the warm earth. Then the boy dozed off and dreamed that the wrestling continued. When he awoke, the night air was cold and the stranger was gone.

On the following day the man returned, this time in the morning. Again they spoke of many things, again they fought, again they rested together and again the youth woke from sleep to find his partner gone. But he was pleased when he reflected on his efforts, for in

that day's wrestling he had felt himself more able to match his opponent's great energy and skill.

The next morning the pattern was repeated. The games continued with no decisive victor—although the stranger seemed somewhat weaker than before— and the day ended with the two combatants lying down together on the grass.

But this time, when the boy woke from sleep, his companion was still lying beside him. The man observed that the next day was the last that they would spend together, for he knew that the prescribed period of fasting and retreat would then be over. They agreed to meet for one final wrestling match.

In this last game, said the stranger, the contest would be unequal. The young Iroquois would defeat him in every encounter and at the end of the match the older man would die.

The youth stared at him in disbelief. Then he folded his arms, shook his head and refused to fight. Wrestling was a test of skill, not a means of murder. And how could he kill someone he had come to respect and love?

The man rose, drew his robe about him and told his young friend to do exactly as he directed, without protest or regret. After their final, fatal game, the boy was to dig a shallow grave. Then he was to divest the man of his headdress and robe and bury him. Finally he was to go home to his people without speaking a word of what had passed, and he was not to mourn. But it would please him, the stranger said, if the youth sometimes found time to visit the grave and sprinkle

The green and golden plants shooting out of the earth changed forever the lives of those peoples who had only known how to kill or forage for their food.

water over the earth that covered it.

Eyes bright with unshed tears, the boy insisted that he would rather die himself than kill his friend. His companion explained that his death was a necessary sacrifice and good would come of it. They would see each other again. Then he disappeared with the day's dying light.

In the morning, the visitor returned and announced that it was time for their final game to begin. The youth hesitated, torn between love and fear, then flexed his muscles and attacked.

Their wrestling that day was hard and furious. The sky wheeled above them, the birds fell silent, the sun stood still. The man battled fiercely as if oblivious of his own foretold fate, and the sweating, straining boy felt that he, rather than his opponent, would soon have the life drained out of him.

But what had been predicted finally came to pass, and after a final clash of muscle, sinew, bone and will, the game was done. The golden-crowned stranger lay dead upon the grass.

The young brave, no longer a boy, dug the grave and gently laid his companion to rest. Then he went back to his people and puzzled them with his silence. When they asked what spirit message had been vouchsafed to him, he lowered his eyes and said nothing.

His rites of passage completed, he became a hunter, painfully learning those skills upon which the tribe's survival depended. But no matter how far he walked or how weary he felt, he made time to bring water to the burial mound in the woods. Sometimes he looked about him in the vain hope of seeing a familiar, tall figure emerging from the trees. But no one came near except the birds overhead and the small creatures rustling and scuttling on the forest floor.

The time came when his fidelity was rewarded. One hot afternoon, when he brought water for the thirsty earth, something tall and green and golden-crowned rose up from his friend's grave. The prayer that the boy had made in his time of fasting was answered. He had accomplished something that would benefit his entire tribe: He had brought his hungry people the gift of maize ☆

Noah's Miraculous Voyage

The Great Flood that nearly destroyed all living things haunted the memory of the human race. The ancient Babylonians, the Chinese, the Hebrews, the peoples of India, all had their versions of the cataclysm.

In the lands that took their teachings from the Bible, Noah was the hero of the tale. But Europe's storytellers remembered incidents that never appeared in the scriptural account.

They lamented the destruction of certain strange, exotic species that never reached the safety of the ark, either because they lived in such remote places that they failed to hear Noah's summons, or because they were foolish and declined the invitation when it came. Thus the wonderful unicorn—that snowy, horse-shaped creature with a horn on its brow; the redoubtable griffin—leonine and eagle-winged; and the shaggy-coated mastodon were lost to us for all time.

For these and their ilk, the Flood was an end, not a beginning. But for the animals that were taken aboard, paired off by gender, the voyage in that stout ship of gopherwood, three hundred cubits long, would be a journey of hope.

The Devil was not well-pleased with Noah. If it had been up to the Evil One, all of God's creatures could have drowned in the rising floodwaters, and good riddance to them. So he set out to sabotage the ark.

Swiftly shifting his shape, the Devil contrived to enter the craft under the benign gaze of Noah himself. How could the old man have known that one of the tiny, timorous mice scuttling gratefully up the gangplank, darting between the raindrops, was none other than the unspeakable Prince of Darkness in disguise?

For forty days and forty nights, as the chroniclers tell us, it rained, and for one hundred and fifty days more the ark floated on the waters that had covered all the earth. In their tiny cabin, Noah, his wife and family prayed for salvation. The animals huddled in their cramped quarters below deck, taking warmth and comfort from each other, and obeying Noah's command that they should dwell together in perfect peace. Even those who in the wild were enemies should lie down quietly, side by side, for the duration of the voyage.

But behind one of the sacks of grain that Noah had laid by as provender, the mouse who was not a mouse was quietly busy. Unseen and unheard, he began to chew at the timbers of the ark.

Day after day the Devil sucked at the pitch and nibbled away at the gopher-wood. A steadily growing pile of wood-shavings collected around him. Patiently, he bored his way through the hull. He knew that when the hole grew big enough, a great torrent of water would come gushing in, and the ark would sink.

But one day, Noah, tending the beasts in the hold, heard a tiny gnawing sound. He followed the noise to a remote corner, and heaved aside the sacks of grain. There he saw the diabolical mouse, up to his gray haunches in sawdust. Water dripped through the well-chewed spot in the hull.

Furious, Noah pulled off his thick fur glove and tossed it angrily at the rodent. As it flew through the air, the gauntlet, touched by the hand of God, turned into a fur-covered, four-legged, long-tailed creature: the world's first cat. With a piercing meow, the feline leaped upon the mouse and gobbled him up.

Noah was grateful that the cat had saved the craft from destruction, but he was well aware that she had also violated the supreme law of the ark, by which all animals were enjoined to live peacefully together without bloodshed. So Noah had no alternative but to carry the furry creature up on deck and drop her over the side, into the churning waters.

But God, who had created the cat, did not let her drown. He showed the animal how to hold her whiskered head above the water and swim. Then God spoke to Noah and told him that the gnawing mouse was no ordinary vermin, but the Devil himself in disguise. The cat should be spared.

So Noah threw out a rope and hauled the waterlogged beast back on board. After shaking herself vigorously, the cat stepped elegantly across the deck and found a sunny patch, where she lay down to dry her fur. From that moment on, every cat that was ever born would hate water, but would hunt for mice in hopes of catching the Devil. And cats' eyes would gleam uncannily in the dark, because all cats have a little bit of Satan in them.

After this narrow escape from disaster, the hole in the hull of the ark had to be plugged. One of the snakes offered to plug the gap temporarily with his triangular head, until Noah and his sons could prepare a new piece of timber to replace the damaged plank.

When the work was done, and the snake was relieved of his post, he demanded payment for his efforts. When Noah asked him what

reward he expected, the serpent announced that he had never tasted food so sweet as the blood of human beings, and one of Noah's sons looked particularly tasty. Surely Noah could spare him; he had other sons, so it would be an easy sacrifice.

Noah refused outright. And when he saw the menancing glint in the reptile's tiny eyes, and saw the twitching of his poison-laden fangs, he raised his staff and struck the snake a lethal blow. Then, for he knew not what sinister powers this mysterious creature had, he cut the snake into tiny pieces and threw them into the fire that burned all night in a brazier, lighting the deck.

When the snake was thoroughly incinerated, he gathered up the ashes and threw them out to sea. But the wind blew them back again, transformed into gnats and mosquitoes and horseflies and midges, and all the other pestilential insects that suck the blood of beast and man. So the snake got his wish after all.

Soon after this event, the birds on board the vessel grew very restless. The chattering magpies, who had been a noisy plague to all their traveling companions, refused to remain on the ark any longer. Instead they swooped over the waters, exclaiming over the floating corpses of the flood victims. And from that day forward, say the storytellers, magpies have been feared as an omen of bad luck.

The kingfishers also left the ark, but at Noah's bidding, to see if the waters had abated. The birds soared upward into the heavens, taking on forever the red of the sun and the glorious blue color of the sky. But they brought back no news of dry land.

Next, Noah sent out a raven, but the bird never returned, preferring to prey upon the floating bodies of the dead. As a mark of

God's displeasure, it was forced to live off carrion forevermore.

Finally, Noah released a dove into the air. In time, the bird came back to the ark bearing an olive leaf, and Noah rejoiced at this sign that the floodwaters were, somewhere and somehow, abating. As a reward for bringing the good news, God gave the dove its pure white sheen, and a coat of plumage that never molts.

Noah steered the ark in the direction from which the dove had come. He found dry land thrusting up from the waters, the high peak of the mountain that is now called Ararat.

At long last the voyage was over. In pairs, as they had embarked, the birds and beasts and crawling things emerged from the ark and looked about them. Now the world could begin again.

139

Picture Credits

Bibliography

Aldington, Richard and Delano Ames, transl., *New Larousse Encyclopedia of Mythology*. London: The Hamlyn Publishing Group, 1985.*

Allen, Richard Hinckley, *Star Names and Their Meanings*. New York: G.E. Stechert, 1899.

Armstrong, Edward A., *The Folklore of Birds*. Boston: Houghton Mifflin, 1959.

Axtell, James, ed., *The Indian Peoples of Eastern America*. New York, Oxford: Oxford University Press, 1981.

Baring-Gould, Sabine, *Curious Myths of the Middle Ages*. New York: University Books, 1967.

Belting, Natalia, *Cat Tales*. New York: Henry Holt, 1959.

Belting, Natalia, *The Earth Is on a Fish's Back*. New York: Holt, Rinehart and Winston, 1965.

Birch, Cyril, *Chinese Myths and Fantasies*. New York: Henry Z. Walck, 1961.

Brain, Robert, *The Tribal Impulse*. London: Macdonald and Jane's, 1976.*

Brown, W.J., *The Gods Had Wings*. London: Constable, 1936.*

Bulfinch, Thomas, *The Golden Age of Myth and Legend*. London: Harrap & Co., 1919.*

Cathon, Laura E., and Thusnelda Schmidt, *Perhaps and Perchance*. New York: Abingdon Press, 1962.

Cavendish, Richard, ed., *Legends of the World*. London: Orbis, 1982.

Cavendish, Richard, ed., *Man, Myth and Magic*. 11 vols. New York: Marshall Cavendish, 1983.*

Converse, Mrs. Harriet M., *Myths of the New York State Iroquois*. Albany: University of the State of New York, 1908.

Dähnardt, Oskar, *Natursagen*. New York: Burt Franklin, 1970.

Eberhard, Wolfram, *Chinese Fairy Tales and Folk Tales*. London: Kegan Paul, Trench, Trubner & Co., 1937.

Eliade, Mircea, *Patterns in Comparative Religion*. Transl. by Rosemary Sheed. London: Sheed & Ward, 1958.

Elwin, Verrier, *Myths of Middle India*. Oxford: Oxford University Press, Indian Branch, 1949.

Farmer, Penelope, compiled and edited, *Beginnings*. New York: Atheneum, 1979.

Folkard, Richard, Jr., *Plant-Lore, Legends and Lyrics*. London: Sampson Low, Marston, Searle, and Rivington, 1884.

Gallant, Roy A., *The Constellations*. New York: Four Winds Press, 1979.

Gaster, Theodor H., *Thespis: Ritual, Myth and Drama in the Ancient Near East*. New York: Henry Schuman, 1950.

Graves, Robert, *The Greek Myths*. Vols 1. and 2. Harmondsworth, England: Penguin Books, 1985.

Griffis, W.E., *Corea: The Hermit Nation*. London: W.H. Allen, 1882.

Grimm, Jacob and William, *Household Tales*. Transl. and ed. by Margaret Hunt. London: G. Bell & Sons, 1913.*

Grimm, Jacob, *Teutonic Mythology*. Gloucester, Massachusetts: Peter Smith, 1976.*

Hastings, James, *Encyclopaedia of*

Religion and Ethics. Edinburgh: T. and T. Clark, 1908.

Henderson, William, *Notes on the Folklore of the Northern Counties of England and the Borders.* London: Satchell, Peyton and Co., 1879.

Holmberg, Uno, *The Mythology of All Races: Finno-Ugric, Siberian.* Vol. 4. Ed. by John Arnott MacCulloch. New York: Cooper Square, 1964.

Hultkrantz, Ake, *The Religions of the American Indians.* Transl. by Monica Setterwall. Berkeley, Los Angeles, London: University of California Press, 1979.

Ikeda, Hiroko, *A Type and Motif Index of Japanese Folk Literature.* Helsinki: Suomalainen Tiedeakatemia Academia Scientiarum Fennica, 1971.

Jablow, Alta, and Carl Withers, *The Man in the Moon, Sky Tales from Many Lands.* New York: Holt, Rinehart and Winston, 1969.

Kirby, W. F., *The Hero of Esthonia.* London: John C. Nimmo, 1896.

Leach, Maria, ed., *Funk & Wagnalls Standard Dictionary of Folklore, Mythology and Legend.* San Francisco: Harper & Row, 1984.*

Lum, Peter, *The Stars in Our Heaven.* London: Thames and Hudson, 1951.*

Mercantante, Anthony, *Zoo of the Gods.* New York: Harper & Row, 1974.*

Metzger, Berta, *Picture Tales from the Chinese.* New York: Frederick A. Stokes, 1934.

Milne, Mrs. Leslie, *Shans at Home.*

London: John Murray, 1910.

Newall, Venetia, *Discovering the Folklore of Birds and Beasts.* Tring, England: Shire Publications, 1971.

Olcott, Frances Jenkins, *The Wonder Garden.* Boston: Houghton Mifflin, 1919.

Oppert, Ernest, *A Forbidden Land.* London: Sampson Low, Marston, Searle and Rivington, 1880.

Patai, Raphael, Francis Lee Utley, Dov Noy ed., *Studies in Biblical and Jewish Folklore.* Bloomington: Indiana University Press, 1960.*

Power, Rhoda, *How It Happened.* Cambridge: Cambridge University Press, 1930.

Rappoport, Dr. Angelo S., *The Folklore of the Jews.* London: Soncino Press, 1937.

Roy, Sarat Chandra, *The Birhors: A Little-Known Jungle Tribe of Chota Nagpur.* Ranchi: Mission Press, 1925.

Skinner, Charles M., *Myths and Legends of Flowers, Trees, Fruits and Plants.* Philadelphia and London: J. B. Lippincott, undated.

Stone, Merlin, *Ancient Mirrors of Womanhood.* Vols. 1 and 2. New York: New Sibylline Books, 1979.

Thompson, Stith:
Motif Index of Folk Literature. Bloomington: Indiana University Press, 1955.*
Tales of the North American Indians. Cambridge, Massachusetts: Harvard University Press, 1929.

Trigger, Bruce, volume editor, (William C. Sturtevant, general editor), *Handbook of North American Indians.* Vol. 15 (Northeast). Washington: Smithsonian Institution, 1978.*

Underhill, Ruth, *Red Man's Religion.* Chicago and London: The University of Chicago Press, 1965.

Vitaliano, Dorothy B., *Legends of the Earth: Their Geologic Origins.* Bloomington: Indiana University Press, 1973.

Weir, Shelagh, ed., *The Gonds of Central India.* London: The Trustees of the British Museum, 1973.

Werner, Alice:
The Mythology of All Races: African. Vol. 7. Boston: Archaeological Institute of America, Marshall Jones Company, 1925.
Myths and Legends of the Bantu. London: George G. Harrap, 1933.

Williams, C.A.S., *Outlines of Chinese Symbolism and Art Motives.* Rutland, Vermont and Tokyo: Charles E. Tuttle, 1981.*

Zheleznova, Irina Lvovna, compiled, selected and transl., *Folk Tales from Russian Lands.* New York: Dover Publications, 1969.

Zong In-Sob, *Folk Tales from Korea.* London: Routledge & Kegan Paul, 1952.

Titles marked with an asterisk were especially helpful in the preparation of this volume.

Acknowledgments

The editors wish to thank the following persons for their assistance in the preparation of this volume: Windsor Chorlton, London; John Gaisford, London; Oobie Gleysteen, Alexandria, Virginia; Mimi Harrison, Alexandria, Virginia; Rita Hockman, London; Norman Kolpas, Los Angeles; Alan Lothian, Mercatale di Cortona, Italy; John Man, Oxford, England; Robin Olson, London; Birgit Strack, London; Deborah Thompson, London.

Chief Series Consultant

Tristram Potter Coffin, Professor of
English at the University of Pennsylvania, is a leading authority on folklore.
He is the author or editor of numerous
books and more than one hundred articles. His best-known works are *The British Traditional Ballad in North America, The
Old Ball Game, The Book of Christmas Folklore* and *The Female Hero.*

This volume is one of a series that is based
on myths, legends and folk tales.